Heart Healthy Magnesium

Your Nutritional Key to Cardiovascular Wellness

James B. Pierce, Ph.D.

Avery Publishing Group
Garden City Park, New York

This book has been written and published strictly for informational purposes, and in no way should it be used as a substitute for recommendations from your own medical doctor or health-care professional. All the facts in this book came from medical files, clinical journals, scientific publications, personal interviews, published trade books, self-published materials by experts, magazine articles, and the personal-practice experiences of the authorities quoted or sources cited. You should not consider educational material herein to be the practice of medicine or to replace consultation with a physician or other medical practitioner. The author and publisher are providing you with the information so that you can have the knowledge and can choose, at your own risk, to act on that knowledge.

Cover Design: William M. Gonzales
In-house editor: Amy C. Tecklenburg
Typesetter: Bonnie Freid
Printer: Paragon Press, Honesdale, PA

Library of Congress Cataloging-in-Publication Data

Pierce, James B., 1922–
Heart healthy magnesium : your nutritional key to cardiovascular wellness / James B. Pierce
p. cm.
Includes bibliographical references and index.
ISBN 0-89529-579-2
1. Cardiovascular system—Diseases—Nutritional aspects.
2. Magnesium in the body. 3. Magnesium deficiency diseases.
I. Title.
RC669.P495 1994
613.2'8dc20 94-805
CIP

Printed in the United States of America.

10 9 8 7 6 5 4 3 2

Contents

Acknowledgments

I extend my deep gratitude to my wife, Mildred, and to my very good friends Clay and Marjorie Hurst, for critically reading the manuscript and continuously encouraging me to complete it. Clay is a superb copy editor, and his suggestions have greatly improved the quality of the manuscript.

To my colleagues at the University of Massachusetts at Lowell, Dr. William Bannister, Dr. Martin Isaks, and Dr. Ruth Tanner, all of whom are excellent organic chemists and personal friends, I offer my deep gratitude for their valuable comments and suggestions.

I am indebted to the Reverend Dr. Paul E. Toms, my former pastor at Park Street Church in Boston, for directing me to a competent cardiologist and for making valuable and constructive comments about the contents of the manuscript.

Others who have read the manuscript and made valuable comments are my aunt Madge Bruce, my cousins Everett and Jean Wetzel, and my good friends Philip Johnson, Ralph and Mary Milner, Murray and Sara Smith, Charles Tilley, and the Reverend Mac and Libby White.

I owe a special note of gratitude to Brenda Turbeville, Carol Link, and the Friendly Fellowship Class, who graciously arranged seminars for senior citizens interested in magnesium's therapeutic properties. This exchange of ideas was a very positive experience.

As one who had little experience or knowledge concerning cardiovascular disease at the beginning of my encounter with it, I must thank God for providing the inspiration and guidance that led to an amazingly simple solution to my affliction. I am also grateful for all of the events that have led to the successful completion of the manuscript and the publication of this book.

At Avery Publishing Group, managing editor Rudy Shur and his staff have graciously led me through the various stages of publication and offered constructive advice and criticism. Amy Tecklenburg, in her role as copy editor, kindly helped me convert a somewhat ponderous scientific style of writing into an easily readable manuscript. To these, and all the others who contributed to the production of this book, I offer my gratitude.

Preface

Do you get cramps in your legs or back? Perhaps you have a "heavy" feeling in your chest, or unexplained pains in your chest, upper back, or left arm. Maybe you have diabetes or high blood pressure. You may have hurried down a sidewalk once and felt a sharp pain in your leg that made you lame for a week. You could have trouble keeping your hands and feet warm. Perhaps you feel tired all the time. If any of these ailments afflicts you, this book may offer the solution you've been needing for years.

High blood pressure, irregular heartbeats, chest pains, heart attacks, strokes, insomnia, stage fright, fading memory, and stiff, aching, painful muscles can all relate in some way to a deficiency of magnesium in the muscle cells. My discovery of my own magnesium deficiency began with a series of seizures caused by spasms of the coronary arteries. While I was having those distressing experiences, a number of events indicated what caused the spasms and finally led to a simple remedy. It made me so excited to learn this that I feel I must keep sharing it with people.

Unusual as my discovery was, it did not result from any original research I had done, but rather evolved from reading a wide variety of material. From biochemical literature relating to the utilization of the chemical element magnesium by the body, I gained access to the real substance of the cure for my condition.

After getting the initial idea from an article in a newsletter, I began searching through *Chemical Abstracts*, which indexes the chemical literature of the world. Once I had spent several days collecting references to the biochemistry of magnesium, I found that a significant number of them referred to a journal entitled *Magnesium*, with which I was not familiar. This is a relatively new journal that has been published since 1982 (its name has since been changed to *Magnesium and Trace Elements*), and it is not as commonly found in scientific libraries as most other journals are. Perhaps this book will stimulate enough interest to cause more libraries to acquire this valuable publication. I consulted many other scientific and medical journals as references as well in the course of my research.

In the chapters that follow I will share with you what I have learned about magnesium. I wish to give you a body of information that you may take to your physician for his or her consideration in treating your specific cardiovascular condition. This narrative evolves from the experiments, interpretations, and opinions of the experts who conducted the original research and published the information. As you will see, it points coherently to the powerful benefits of having an adequate level of magnesium in our systems. We shall have a front-row seat, as it were, in the laboratories of a number of outstanding scientists. It is far beyond the scope of this book, however, to include a discussion of all the work that scientists have done on the biochemical aspects of magnesium. I apologize to those whose research works are not mentioned.

My goal in writing this book is to share the good news of the curative and health-giving powers of magnesium. The direct benefit for many people will be a higher quality of life, and, in many cases, a longer one. I sincerely believe that I would not be alive today if I had not discovered and taken advantage of magnesium's positive effect on the cardiovascular system.

In my life I have endured many unpleasant and adverse experiences. As I reflect on those events, I am aware that they gave me unique opportunities to learn about life, increase my wisdom, and prepare for the diverse adventures that lay ahead. This book has its roots in one such experience. At first, my future seemed quite uncertain. But the start of this complex ordeal was followed by a sequence of seemingly unrelated events that led to an amazingly simple solution.

Let me share this with you.

Before we begin, I must tell you that I am not a physician. I am a chemist who has spent thirty-four years teaching chemistry at the college level. That adventure gave me a broad base of experience, which I am currently using to assemble and condense the information about the beneficial effects magnesium has on our hearts and circulatory systems.

The introductory chapter of this book tells how I came to know about the life-sustaining properties of magnesium compounds. Center stage is a discussion of the chest pain that led to my hospitalization—how it developed, how my cardiologist treated it, and how a common mineral restored vibrant health. The signal for the cure came from an article in a newsletter that pointed to magnesium as a key to how persons with angina pectoris and other heart-related problems might improve their health.

To understand how something as simple and natural as magnesium can have such extraordinary effects, we must go back to the time when life first appeared on the earth. Chapter 1 examines the essential role magnesium plays in all of life as we know it, its functions in the body, and why most people—including physicians—are unaware of its benefits. In Chapter 2, we will consider why magnesium deficiency is so widespread, how magnesium deficiencies develop, and how an individual might tell if he or she has a magnesium deficiency. I also offer guidelines that will help you to understand how much magnesium you really need.

The following three chapters comprise a discussion of some of the specific health problems that can be related to a magnesium deficiency. Chapter 3 addresses one of the first manifestations of cardiovascular disease, hypertension—what is meant by high blood pressure, how we measure it, what causes it, what some complicating factors are, and what dangers it presents. We also look at the benefits and risks of both conventional and nutritionally oriented treatments for the condition. In Chapter 4, we will look at some of the more advanced stages of heart disease—including chest pains, abnormal heart rhythm, congestive heart disease, and spasms in the arteries—and magnesium's role in relieving and helping to avert these problems. And in Chapter 5, I discuss the so-called "end points" of cardiovascular disease—heart attack and stroke. We will look at what happens during these events, the various things that can lead to them, what the consequences are, and how magnesium may be used even in the midst of such crises to minimize damage to the heart or brain. Here the role of nutritional therapy in helping to prevent heart attacks and strokes comes into prominent focus.

In Chapter 6, we come to the question people inevitably seem to ask: "If magnesium is so good, why did my doctor prescribe aspirin instead?" We will look at some of the scientific studies that have led to the widespread use of aspirin for heart patients, and compare them with the data on the benefits of magnesium. (The results may surprise you.)

The concluding chapter tells you how to put magnesium to work in your own life to improve your health and well-being. It shows how to design and begin implementing a program of nutritional supplementation tailored to your individual needs, as well as how to integrate this program with a healthy approach to diet and stress.

This is the combination that worked for me. And best of all, it is a completely natural approach to heart health. It involves no drugs, and therefore no side effects. Of course, you should consult with your doctor before beginning any health program, including this one—especially if you are currently taking medication for a heart problem, or are suffering from any kind of chronic disease. But I am convinced that many people now taking various medications may not need them, and once they have their magnesium balance restored, they may find—as I did—that nutritional supplementation is more effective at relieving their symptoms than drugs ever were. I hope that by writing this book I will bring to many such people the good news about the health-giving, and potentially life-saving, powers of this amazing element, magnesium.

Introduction: It Happens Without Warning

The temperature was in the twenties, a light snow was falling, and my left jaw was filled with excruciating pain. My watch told me I must endure yet another hour, as I sat with my face jammed up close to the heater duct and continued waiting in my car for the dentist to open his office.

The previous afternoon, he had prepared two teeth to receive a bridge. On each of the two anchor teeth he had installed a temporary aluminum crown. It was inevitable that the pain caused by the electrochemical action between one of the aluminum crowns and the gold inlay on an adjacent tooth would awaken me in the early morning hours and trigger an unscheduled trip to the dentist's office.

When he arrived, a few minutes before nine o'clock, I met him at the door, and I was the first person in the chair. My intense suffering was relieved when he removed the aluminum crowns and replaced them with a pair made of plastic. His only regret seemed to be that he had used aluminum crowns on someone who knew about the relationship between galvanic action and dental pain.

Wide awake, but weary and haggard, I went to work and dragged through the day, wrestling with the usual frustrations in the daily life of a chemistry professor. When I returned home that evening, I was unusually tired from the long day, which had begun at about 3:00 A.M. But my wife insisted that we go to the local YMCA to take a ballroom dancing lesson. I lost that argument, and soon we were dancing the foxtrot and waltz.

During one of the dances, a horrible pain penetrated my throat and chest. It was a sensation unlike anything I had ever experienced before. Even though I sat down, it got worse. The pain was so forbidding that it seemed as though it might have been inflicted by the fiery breath of a mythical dragon or, perhaps, the pitchfork of the devil himself. As my wife and I travelled to the emergency room of the local hospital, the pain gradually diminished until it was gone, but I was truly exhausted. Fortunately, an electrocardiogram showed no abnormalities. Unfortunately, I had no idea what had caused the attack.

About eight years later I developed high blood pressure, for which my physician prescribed a diuretic, Dyazide. Two years after that, the distressing pain in my throat and chest began to recur, and at the most unexpected times—while I was driving to church, lecturing to a class of students, or even sitting at my desk—times when I felt no particular stress. Our family physician of many years diagnosed the affliction as angina pectoris and prescribed nitroglycerin tablets to relieve the discomfort.

Other problems then began to arise. At times I developed a rapid, irregular heartbeat, especially after meals, at which I always drank four or five cups of coffee. Fortunately, that condition was largely alleviated by avoiding caffeine. Overeating caused the same symptoms, but the remedy for that bad habit was obvious.

Then, in the last week of May 1986, as I was working in the lawn, I had a debilitating seizure. That *really* got my attention. The direct result was hospitalization, with its attendant battery of examinations and procedures. The most serious of these was a catheterization of the coronary arteries that feed the heart muscles, to determine whether the blood vessels were blocked.

Amazingly, the cardiologist reported that my coronary arteries had very little, if any, blockage; but he told me that my symptoms were "impressive." I had apparently joined the estimated 10 to 30 percent of patients who display the *symptoms* of diseased arteries, but who show little or no evidence of arterial obstruction. In order to control the anginal attacks and help reduce my blood pressure, the doctor prescribed a calcium-channel-blocking medication called Cardizem. He diagnosed the angina as originating from spasms in the coronary arteries, rather than from blockage of the blood vessels.

For two years I took the medication and my condition was essentially static. Then it began to get worse. I was constantly exhausted. Each day I had to struggle to complete my work. After lecturing for fifty minutes, I had to rest for at least twenty to thirty minutes. I became depressed as my memory and my ability to reason began to fade. Afternoon naps at my desk were a necessity. Obviously, my performance in the classroom was seriously impaired.

During the spring of 1988, the angina occurred more intensely and more frequently than ever, and I developed a severe pain in my lower back and thigh. It made me so lame that I needed the support of a cane to walk. The muscles of my left leg were under such continuous tension that they made a noise resembling the plucking of stretched rubber bands as they moved against one another. They refused to relax.

When I visited the cardiologist in early June 1988, the pain was so bad that I could not take a stress test on the treadmill. Fortunately, my blood pressure was 120 over 70, which led the cardiologist to say, "Don't change anything; keep taking the Cardizem." But even though I was taking the medication, my angina was getting worse. I couldn't recover my strength and vigor. The attacks—which can be fatal—were gaining in frequency and severity, and my health was on a downward spiral.

Then, later that month, in a newsletter called *Privileged Information*, I came across an article describing how a major American study of aspirin's effect on inhibiting heart attacks was flawed.[1] The results of this study, in which half of the 22,071 participants (all physicians) took aspirin every other day and the other half took a placebo, had been widely reported in the press and hailed as offering a dramatically successful method of preventing heart attacks.[2] The first objection cited in the newsletter article was that the participants taking aspirin had taken it in the form of Bufferin, which contains both magnesium and calcium in addition to aspirin. The second objection was that a similar study done in England had shown that aspirin had no statistically significant effect on heart attack risk in the 5,139 physicians who participated.[3]

The American physicians who took Bufferin were reported, in two different accounts of the same research, to have had 47 percent and 44 percent fewer heart attacks than the physicians who consumed a placebo. In the English study, the participants ingested pure aspirin every day, and they received more than three times as much aspirin per week, month, and year as the Americans, with no demonstrable benefit. Therefore, the author of the newsletter article proposed, the mineral content of Bufferin—not the aspirin—may have been the beneficial factor that lowered the risk of heart attack among the American physicians.[4]

After reading this, I began taking dolomite (calcium magnesium carbonate) tablets. Relief from my back and leg pain occurred in a matter of days, and by the end of November my cardiac condition had so improved that I decided, on my own, to gradually discontinue taking the Cardizem. Angina was no longer a problem. My strength and vigor were returning, although they were not yet back to normal. Two weeks after I stopped taking the heart medication, my cardiologist found it difficult to believe the improvement he saw in my electrocardiogram. On his advice I returned to the Cardizem treatment, but only for a short time.

From all indications, my problems had been caused by a deficiency of magnesium in the cells of both the cardiovascular and skeletal muscles, with a possible deficiency of calcium, potassium, and other minerals throughout my system.

Because the body is so complex, I could not say that the beneficial effects resulted exclusively from the ingestion of magnesium; I had taken both calcium and magnesium in the form of dolomite tablets. These two elements are chemically similar, but the body uses them in different ways. Magnesium resides and functions principally within the cells, whereas calcium's main functions occur outside the cell walls. For the body to function normally, both calcium and magnesium have to play their appointed roles, with calcium and magnesium competing with each other on opposite sides of the cell walls. Both are necessary to maintain proper muscular tone, contraction, and relaxation.

Professor Bo Siesjö of the Laboratory for Experimental Brain Research at Lund University Hospital in Lund, Sweden, advises that if too much calcium moves into the cells, it displaces some of the magnesium that belongs there, and causes the cells first to contract, and then to die.[5] By taking tablets of calcium magnesium carbonate, I was receiving both elements as they occur in nature, in the form of limestone known as dolomite.

To learn about the effect and possible beneficial influence of the calcium in the dolomite, we need to consider that magnesium exerts a regulatory influence over the actions of calcium, potassium, and sodium. During the time I was seriously deficient in magnesium, my calcium level was also below normal, no matter how much calcium I consumed. In view of this, plus other information that we shall consider in later chapters, I have no doubt that magnesium was the key nutrient that led to my recovery.

In the spring of 1989, my family physician advised me to discontinue using Dyazide, the diuretic medication. My blood pressure was still 120 over 70, even in the absence of the heart medication, which I had discontinued earlier. Discontinuing the diuretic had no effect on my blood pressure either. (I would add that my blood pressure has remained normal since I began taking the dolomite tablets.)

All prescription medications are legalized laboratory chemicals that should be taken only as long as the benefits exceed the risks. With some medications, the risks exceed the benefits before you even begin taking them. Typical risks become evident in the form of side effects, such as dizziness, nausea, headaches, rashes, and the like. Some risks may be more subtle, and/or more dangerous. In some cases, drugs can initiate other conditions that are worse than the ailments they are prescribed to treat. *The Pill Book* from Bantam Books lists the side effects of many different medications.[6] Looking through its entries can be scary reading.

With Cardizem and Dyazide I didn't *seem* to experience side effects, but in reality I did—it was just that they were subtle. During the time I was taking these two medications, my memory and reasoning capabilities gradually deteriorated to the point that I began having difficulty in the classroom. It was terribly embarrassing to be unable to recall facts and procedures that I had been teaching for years, or even the names of good friends whom I encountered frequently. This and several other conditions led me to retire from teaching in the spring of 1989. I am truly grateful that those side effects were reversible, and that my mental capacity has returned to normal.

To sum up, then, after two years of taking Cardizem, I still experienced unpredictable seizures of angina, and they were occurring more frequently and more intensely than ever. Then I encountered the suggestion in a newsletter article that magnesium was effective in reducing high blood pressure and other cardiovascular ailments, which led me to begin taking dolomite tablets. I disobeyed my cardiologist and stopped taking Cardizem. This is one of the standard medications for heart patients, but in my case it had, at best, only minimal beneficial effects.

I had aged rapidly during my bout with angina. I had become essentially an invalid. My wife mowed the lawn and did all of the tasks that required even light physical exertion. But by January of 1989—about seven months after I began taking the dolomite—I felt twenty years younger. My vitality was returning. In the spring of that year, I began resuming all of the physical chores that had been denied me by the heart spasms and the resultant lack of oxygen flowing to the heart. By the end of 1989, I felt better than I had for many years.

As time went by, I gradually increased my dosage of dolomite to the equivalent of 667 milligrams of magnesium per day, and I felt great. In fact, I could not remember having ever felt better. Then, in my euphoria, I began to forget to take the dolomite and other dietary supplements. Unfortunately, nature can be very unforgiving and constant in the enforcement of its laws, and we must always pay for our negligence. On this occasion, it could very well have cost me my life.

On February 1, 1990, while in the medical library at the Shands Hospital in Gainesville, Florida, searching the chemical and medical literature for research results that could be used to support the idea that magnesium is beneficial in the treatment of spasms, I had the most unusual seizure I had ever experienced. The familiar discomfort was accompanied, for the first time, by nausea, profuse perspiration, and blurred vision. I nearly lost consciousness as I cradled my head in my arms on the desk.

This is a set of symptoms that can indicate the beginning of a heart attack. My world had fallen apart! I had discovered something great, and now it seemed it no longer had the value I had imagined. In fact, the seizure was so severe that I thought I had seen my last ray of sunshine.

At the onset of the seizure I took two nitroglycerin tablets and 250 milligrams of magnesium. When I recovered sufficient strength to walk, I slowly made my way to the emergency room. There a very competent intern examined me and recommended that I be admitted to the hospital. After admission I had the usual electrocardiogram and blood tests. My heart function was monitored continuously until the next morning.

Shands is a teaching hospital, associated with the University of Florida, which made it different from any other hospital I had ever been in. I was placed under the care of a team of young physicians and medical students, headed by a senior professor and resident physician who was a cardiologist. This team reviewed my medical history. They even obtained the reports of my heart catheterization and other medical tests from Boston.

After considering all of the data, they reached the surprising conclusion that my spasms were not necessarily cardiac related, but could have occurred in the esophagus. Since spasms of the esophagus and the coronary arteries may be interrelated, it is difficult to distinguish one from the other. They are treated by the same medications. (In my case, however, the loss of strength and near loss of consciousness I experienced strongly suggest that the spasms occurred in the coronary arteries, seriously decreasing my heart's functions.)

I was released the next day. The recommended treatment was that I continue taking the same minerals and vitamins I had been using, but to make certain I took them on a regular schedule. No medications were prescribed.

As we shall see, whether the spasms were in the coronary arteries or in the arteries and muscles of the esophagus was really not important to the cure. A deficiency of magnesium can cause spasms in any of a number of different arteries. Spasms can occur in the arteries of the fingers and toes, the esophagus, the heart, the rectum, or even the brain. They can occur individually or in any combination.

Contrary to my worries about the impact of this seizure, it ended up providing a tremendous endorsement for nutritional therapy. The professional team of physicians at Shands recognized magnesium as an agent for soothing the smooth muscles of the cardiovascular system. Thus, an event that I thought would disprove the usefulness of magnesium for treating spasms resulted in wholehearted approval by a respected cardiologist and his staff. It also underscored the necessity of taking the supplements regularly.

As I thought about these developments, I didn't know exactly what had caused the seizure. I guessed that my failure to take the magnesium supplements on a regular basis had placed me in a temporary state of magnesium depletion. Since I was still unfamiliar with the way magnesium works and had no way of knowing whether I was taking too much or too little, I established a regular dosage of 333 milligrams per day. This was close to the recommended daily allowance (RDA) for men, 350 milligrams, and it was in addition to the magnesium in my diet. For myself, a man weighing 215 pounds, I regarded the RDA as the minimum amount I should consume in magnesium supplements each day. And I made certain that I took them regularly.

I felt quite well—until I encountered the stress of driving my car all day long for several days. After three consecutive days of driving on interstate highways, I needed about a week to recover my strength and reduce my sleep time to normal. Then, on a similar trip, I took 250 additional milligrams of magnesium at about 2:00 P.M. each day, and the results were remarkable. With this increased dosage, I was able to drive all day long for three consecutive days and not even feel tired in the evening. Perhaps truck drivers who pilot those heavy rigs on interstate highways could enhance their alertness and health by judiciously taking magnesium on a regular basis. In fact, everyone who lives or works in a stressful environment would probably do well to heed this advice.

From my earliest experiences with taking magnesium as a supplement I kept asking myself the question, Is magnesium therapy *really* beneficial, or did something else cause the miraculous recovery I have experienced? As I explored the chemical, biological, and medical literature, I found the answer to be a resounding endorsement for magnesium. In fact, magnesium therapy is even more powerful and beneficial than I could have imagined. I invite you to look with me at some of the wonderful things magnesium does in regulating the operation of our cardiovascular systems.

As a preview, consider the report of Drs. R.S. Parsons, T.C. Butler, and E.P. Sellars in the British medical journal *Lancet*. These physicians tell us that over a period of four years, they administered magnesium sulfate (Epsom salts) by intravenous injection to many patients who had diseased hearts and blood vessels and who were also deficient in magnesium. All patients showed striking improvement, in the form of freedom from chest pain, absence of recurring heart attacks, increased ability to exercise without becoming fatigued, and the return of electrocardiograms to normal.[7] This is impressive! But it is only the beginning. Let us now turn our attention to the many positive ways in which magnesium affects our lives.

1

Magnesium's Role in Our Lives

Wouldn't it be wonderful if there were a single natural nutrient that could:

- help to control blood pressure;
- soothe muscles and prevent cramps;
- suppress the formation of kidney stones;
- inhibit the abnormal deposition of platelets and formation of clots in the blood vessels;
- increase blood flow without increasing blood pressure;
- help to normalize nerve impulses and keep the heart beating with a normal rhythm;
- prevent spasms of the muscles and blood vessels;
- protect muscle cells from being injured by an abnormal infusion of calcium;
- assure the normal operation of the body's cells by controlling the correct balance of electrolytes;
- reduce the risk of angina; and
- greatly reduce the risk of heart attack or stroke?

This may sound like an advertisement for a universal medicine, but there is such an agent in our bodies. It is magnesium, the element that does all of these things—and more.

WHY IS MAGNESIUM SO IMPORTANT?

Although we may not think about it much, we are closely connected to our environment. The composition of our biological systems is intimately related to that of our natural habitat—including the air, the water, the land, and all the components of our food supply.

Long before any life existed on this planet, all of the earth's chemical elements, or their precursors, existed in a hot primordial soup comprised of molten lava under a dense noxious vapor. A relatively small amount of matter came from outer space in the form of meteorites, cosmic radiation, and possibly even asteroids that fell to earth, but the bulk of the earth's components have always been here. The earth cooled, land masses formed, and the vapors condensed to form liquids and solids composed of chemical compounds familiar to us today.

Fossil records tell us that many millions of years before the atmosphere contained oxygen or humans appeared on the earth, single-celled species—which fed on sulfur dioxide, sulfate minerals, carbon dioxide, and other seemingly unlikely chemicals—were abundant in the oceans. Strange as it may seem to us, they didn't use the oxygen in these materials. They excreted it.

That living cells can exist without oxygen may seem surprising, but even today there are bacteria that live in anaerobic environments (habitats without oxygen) and that are actually poisoned by oxygen. In garbage landfills, anaerobic bacteria thrive and produce large quantities of methane gas.[1] Methane is the major component of natural gas. In India, people use methane produced by the action of anaerobic bacteria on organic waste to cook their food.

Over a span of many millions of years, the waste product of these early cells—our life-sustaining oxygen—reacted chemically with the iron dissolved in the oceans and made it precipitate (separate out as a solid) as insoluble iron oxide.[2] Not only did this give us some very rich deposits of iron oxide ores, but it reduced the iron content of the oceans to a very low concentration. Carbon dioxide—a product of volcanic action—was abundant in the oceans, and it reacted with dissolved calcium to form calcium carbonate, which also is insoluble. (Calcium carbonate is the principal component of limestone, marble, and seashells.)

Oxygen then saturated the oceans and began to accumulate in the atmosphere. Cells that fed on oxygen began to develop and, over the course of many millions of years, gradually became present in great abundance.[3] This is an example of the chemical adaptability and development of life forms. A modern example of this adaptability is the development of strains of insects that are resistant to insecticides. Life is persistent.

The developing oxygen-breathing cells required an element to serve as a regulator of chemical reactions, and of all the elements that could perform this function, magnesium remained dissolved in the oceans in the highest concentration. Magnesium's unique chemical properties also qualified it to regulate the concentrations of other electrolytes and control the reactions of enzymes. Consequently, as their structures and essential functions took shape, the emerging life forms incorporated magnesium into their cellular architecture and metabolisms in many different ways. This characteristic then carried over into the more complex organisms when cells became more specialized and merged to form them.

Living species proliferated, and cells began to merge. Cells became specialized and formed organisms. New animal species used oxygen to extract energy from their food supply. Plants used carbon dioxide and other waste products from animal metabolism, as well as decaying animal carcasses, as food. Animals consumed the plants, their fruit, their seeds, and their excreted oxygen to support life. This balance of natural processes has sustained life for millions of years.

Animal species obtain their food supply by consuming other species. In order to have consistency in the food chain, all living creatures must (and do) have similarities in their chemistries. As a result, all vertebrates and invertebrates have a characteristic dependence upon magnesium for the health and functioning of their cells. They cannot live without it.

Plants require magnesium as a vital component of their metabolic chemistry as well. In fact, every molecule of chlorophyll (the green pigment plants use to produce their food through photosynthesis) has at its center an atom of magnesium. The beautiful green of our vegetated landscape thus depends on the presence and chemical action of magnesium. And photosynthesis, which depends on the presence of magnesium, also produces our life-sustaining oxygen. Magnesium is everywhere. It is the eighth most abundant chemical element in the earth's crust. Without magnesium, life as we know it would not exist.

MAGNESIUM'S ROLE IN THE BODY

Whether you believe that humans evolved or were created, or that a combination of the two brought us forth to walk on this earth, we are an integral part of our environment. When compared with the age of the earth, mankind's origin is relatively recent. We are closely allied with the earth, and with the plants and animals that are here with us. We use them as food to sustain our bodies. Their substance is compatible with ours. We breathe air to get the oxygen necessary to oxidize the food in our bodies (the equivalent of burning fuel) and supply us with life-sustaining energy. We maintain our bodies' fluids by drinking water, so abundantly supplied all around us.

Potassium and magnesium are the most abundant metallic elements in our cells, and in the cells of all other organisms. Their special properties regulate the delicate balance of chemical processes necessary to life. Although potassium is the more abundant of the two elements in the cells, magnesium is dominant in regulating life's essential biochemical processes.

Calcium also plays an important role in human chemistry. Magnesium and calcium have similar chemical properties, but each works in a unique way in the body. Both are present in the blood, the muscles, and the bones, in the form of charged atoms called ions. Of the two elements, magnesium predominates in the muscle cells. It tends to control and balance the concentrations of other mineral ions, such as calcium, potassium, and sodium.

The molecules needed to make protein, starches, glycogen, and many other components of the body and food chain depend upon the special ability of carbon to form large, stable molecules, arranged in very specific configurations. Other atoms, such as oxygen, nitrogen, sulfur, and phosphorus, are included in these configurations, situated at definite locations within the molecules. All of these organic compounds have very specific arrangements of their component atoms, and the molecules that result have unique chemical properties that depend on their geometric structures and compositions.

Magnesium is essential to the operation of over 300 enzymes that cause many life-sustaining biochemical reactions to take place. To activate the enzymes used to make protein for our muscles, hair, and fingernails, our bodies need magnesium. Magnesium is also a key factor in producing the nucleic acids (DNA and RNA) that contain the hereditary information in the nuclei of our cells. The energy for muscular motion comes from oxidizing glucose through a series of biochemical reactions in which magnesium is an essential participant. Magnesium is a key actor in transmitting nerve impulses and maintaining the integrity of cell walls. It also regulates the passage of calcium across cell boundaries. Dr. Sheldon Saul Hendler tells us that magnesium plays an essential part in the operation of all major biochemical processes and that it is absolutely essential to life.[4] No wonder it has such a profound effect on cardiovascular health.

Because of the close alignment of nature, all vertebrates and invertebrates depend on magnesium to maintain the health and functioning of their component cells. In the mitochondria of the cells (the sites where glucose is oxidized), magnesium is chemically coupled with a substance called adenosine triphosphate (ATP). This positions magnesium as a catalytic agent at the seat of energy generation. The process is complex, but the net effect is that muscular motion is powered by the energy released when phosphoric acid molecules split off from ATP, leaving adenosine diphosphate (ADP) as a product of the reaction.

Of course, since energy is released to the cell when ATP converts to ADP, the reverse reaction—restoration of ATP from ADP—requires the input of energy. The source of that energy is the oxidation of glucose through a complex series of biochemical reactions. This conversion of ATP to ADP, and of ADP back into ATP, is the basic process that supplies the cells with the energy they need to perform all of their appointed functions, and it cannot take place without the presence of magnesium. And this process, which produces muscular energy, is essential in the operation of our hearts and vascular systems.

Magnesium is a natural calcium-channel-blocking agent. That is, it protects cells from abnormal infusions of calcium, which, in severe instances, can cause cells to die.[5] The amount of calcium inside the cells relates directly to a muscle's tone, or how tense the muscle is. By displacing and excluding excess calcium from the space inside the cell's membrane, magnesium functions as a muscle relaxant. Magnesium's role in relaxing the muscles translates into such benefits as reducing and controlling blood pressure and relieving excessive muscular tone. Thus, it effectively relieves muscle spasms in both skeletal and cardiovascular muscles. This makes magnesium a real lifesaver.

Magnesium also inhibits the deposition of blood platelets. Yet in spite of this antiplatelet activity, and the ability to enhance the flow of blood in all blood vessels, magnesium has no tendency to cause hemorrhage. This is in sharp contrast to the usual antiplatelet drugs prescribed to prevent the formation of dangerous clots in the blood vessels. For example, aspirin can cause bleeding in the stomach and raises the risk of hemorrhage in small blood vessels throughout the body. Because older people are at higher than normal risk of having a hemorrhagic stroke (a stroke caused by the rupture, rather than the blockage, of a blood vessel), it is risky for them to take aspirin to prevent blood clots. The physicians who took aspirin in the Ongoing Physicians' Health Study had twice as many hemorrhagic strokes as those who took a placebo.[6]

Perhaps the most effective, and hazardous, anticoagulant is warfarin. This is a compound used in rat poison to induce lethal hemorrhages. In reduced dosages, it is prescribed by doctors to prevent or dissolve blood clots. It is sold under the brand names Coumadin, Sofarin, and Panwarfin. Needless to say, these are powerful and dangerous medications.

In view of what magnesium can do, it seems appropriate to suggest that people who are now taking antiplatelet drugs should consult with physicians who can advise them instead as to how they can maintain their balances of magnesium, potassium, and vitamins A, C, and E at normal levels. If an individual's condition permits making this change, he or she would be freed of the expense—and the risk—of taking the prescription medications. We need some honest, well-designed studies to clarify this issue.

WHY HAVEN'T WE HEARD ABOUT MAGNESIUM?

You may wonder why the general public is unaware of magnesium's importance to health. There are many reasons, including the following:

- Scientists and physicians, like other people, tend to resist change and new ideas.
- Nutritional supplements are not patentable. They are therefore not as profitable for pharmaceutical companies to make and sell as the proprietary drugs are.
- The profit motive penetrates deeply into the administration of health care.
- Physicians are constantly bombarded with advertisements from pharmaceutical companies that extol the merits of new and current drug treatments.
- Physicians are trained to identify physical problems and prescribe drugs to solve them. Many pay more attention to alleviating symptoms than to eliminating the underlying causes of disease.
- Medical schools to a large extent ignore—and may even discourage—the practice of nutritional therapy. Many do not require even a single course in nutrition.
- Physicians who do not conform to established procedures and approaches risk being ostracized and perhaps even losing their professional privileges. This pressure can be intense.
- Scientific reports must be explained and translated into language that most people can understand before the information they contain can be appreciated. Often the information is out there, but in forms that the average person is unlikely to encounter.

In fairness, we must recognize that doctors' reluctance to try new therapies is not based entirely on prejudice or peer pressure. The scientific community recognizes its responsibility to protect the public from radical new ideas that could be more harmful than beneficial. You may recall instances in which these safeguards were ignored, resulting in disaster. One example that comes to mind is the case of thalidomide, a sedative that was used in Europe to treat pregnant women. The use of this drug resulted in many babies being born without arms or legs, or with stubs of limbs that had limited function.

Magnesium, however, is a completely natural element that is *required* by the body. By controlling the concentrations and activities of specific electrolytes (dissolved forms of essential minerals) in the body's fluids, it tends to normalize the transmission of nerve impulses. Major benefits from this are the maintenance of a normal heart rhythm and the efficient operation of thought processes. In the blood vessels, magnesium helps to control the deposition of blood platelets and inhibit the formation of unnecessary clots. Over 300 enzymes—catalysts that cause specific necessary biochemical reactions to occur in the body—need magnesium in order to do their work.

Muscular motion could not occur without the chemical action of magnesium. Magnesium plays a key role in the generation of energy from glucose. In balance with calcium, it causes muscles to relax and contract properly. The heart muscles and the smooth muscles in the blood vessels are particularly sensitive to the balance of magnesium and calcium, and depend on them for the control of normal blood pressure and heartbeat.

In short, magnesium is involved in the most basic processes that make life as we know it possible. As we shall see in the following chapter, the major concern most of us should have about magnesium is that we may not have enough of it in our bodies to maintain optimum health.

2

Why We Have Magnesium Deficiencies

Of the 248,709,873 people counted in the 1990 United States Census, 70 million suffer from some kind of cardiovascular ailment. That's 28 percent of our total population! One major reason for this is that most people in industrialized societies consume less than the recommended daily allowance of magnesium and other essential nutrients. This is a sad state of affairs—sad because the incidence of heart and circulatory diseases increases as the population's consumption of magnesium decreases.[1] In fact, Heljä Pitkänen of the Salco Research Center in Helsinki, Finland, tells us, "Processing of foodstuffs by industry adds sodium and removes virtually all of the magnesium. It is thus possible that today's greatest dietary hazard is the universally low level of magnesium intake."[2]

We commonly think that people who eat balanced meals are getting all of the nutrients they need to sustain their bodies in a state of vibrant good health. For many of us, however, this is becoming increasingly difficult, especially as we grow older. There are a number of reasons for this:

- We eat too much processed food.
- The absorption of nutrients in the digestive tract is not 100-percent efficient. Many of us absorb smaller amounts of the nutri-

ents we eat than the body requires. And as we age, the efficiency of nutrient absorption seems to decrease.

- Synthetic fertilizers, now commonly used in agriculture, fail to replace all of the nutrients that plants take from the soil.
- Industry causes serious pollution of the air, water, and the food supply. This increases our nutritional need for essential vitamins and minerals. For example, the chlorine commonly added to drinking water uses up vitamins C and E, and the purification of water depletes it of calcium, magnesium, iron, potassium, zinc, and other minerals.
- As people age, they eat less, so they get less of the essential vitamins and minerals from their diets.
- The stress of life in our modern society causes us to excrete vitamins and minerals at an increased rate. Thus, less of these essential nutrients remain in our bodies long enough for them to perform their appointed functions.
- Some medications cause vitamins and minerals to be wasted. Serious offenders in this category are diuretics, which are frequently prescribed to regulate blood pressure.

Any of these factors, among others, can lead to nutritional deficiencies that can cause you health problems. Then, when you present your physical complaints to a physician, who sees you for only a few minutes, he or she is likely to prescribe one or more medications to counteract your symptoms. But medications cannot cure nutritional deficiencies. They can only make you feel worse than you already do.

For example, if a person has high blood pressure, which can result from a deficiency of magnesium, a common treatment is to prescribe a diuretic. But diuretics cause the elimination (through the kidneys) of the very magnesium and potassium that are needed to maintain normal blood pressure and the general good health of the cardiovascular system.

HOW MAGNESIUM DEFICIENCIES DEVELOP

There are many different things that can cause a person to become deficient in magnesium, from insufficient intake to complex metabolic disorders. But of all the causes of deficiency, an inadequate diet is one of the most common. (Fortunately, as we shall see later in this book, it is also one of the easiest to correct.)

Poor Diet

Many people do not have enough magnesium in their bodies simply because the foods they eat fail to provide it in sufficient quantities. A major reason for this is that we eat so many processed foods. Food companies refine their products to make them more appealing and to increase their shelf life. Then, because the nutrients in the food have been so severely diminished, they "enrich" them with vitamins and minerals; but this fails to restore the original nutritional quality. Common examples of this are breads and cereals. Our culture's dependence on refined foods is a major reason why many of us do not get enough nutrients from food alone. Heljä Pitkänen, who warns of the hazards of magnesium deficiency, cites the industrial processing of food as a principal cause.[3] In the refining of grains, vegetables, fats, and oils, the losses of magnesium range from about 70 to over 90 percent.

The food distribution systems of industrialized nations are efficient means of feeding vast urban populations. But as companies market processed foods to the general population, they propagate a general nutritional deficiency and place the masses at risk of developing an array of degenerative diseases. These include high blood pressure, abnormal heart rhythm, plaque formation in the arteries, muscular and cardiovascular spasms, strokes, and complications from diabetes.

Dietary deficiencies come in a wide range of severities. Many people have borderline deficits that cause them to be just slightly less healthy or productive than they would like to be. Others have full-blown diseases caused by improper diet. And in between are many folks who are in various stages of becoming the victims of bad nutritional habits.

You may not be able to avoid diseases caused by viruses and other microorganisms, but you can avoid nutritional diseases that develop as the result of an unbalanced diet. For example, it is a fairly safe assumption that a liquid high-protein diet or other extreme dietary plan is likely to be lacking in vitamins and minerals. All low-calorie diets should be suspected of being deficient in essential nutrients and should be supplemented with essential vitamins and minerals. The caloric content of the supplements is negligible.

There is one ready source of dietary magnesium and other minerals that may help you avoid deficiency. If you live in a hard-water area, check to see if magnesium is one of the principal minerals in the water. You may be very fortunate. A number of studies done in different countries show that the incidence of cardiovascular disease is greater in areas where the water has a low magnesium content.[4] The link between cardiovascular disease and a lack of magnesium in drinking water was first proposed by Dr. R.S. Parsons of Australia and his coworkers.[5] In the United States, magnesium-rich water supplies occur mostly in the central part of the country, in a band including Texas and the states to its north. The eastern and western states have water supplies that, in general, are more deficient in magnesium. And as one might expect, the incidence of cardiovascular disease in those states is higher than it is in the central part of the country.[6]

Even if your diet or your drinking water provides abundant magnesium, however, there are other factors that can cause a magnesium deficiency. Some of these come upon us quickly and develop without warning.

Stress

Stress is a subtle cause of magnesium deficiency. Stress can be inflicted by many different things, both physical and psychological. Frustrating situations that we can't control are a major cause of stress. This can mean getting stuck in rush-hour traffic, being faced with an unfamiliar task, encountering an abusive person, or visiting the dentist's office. These are obvious sources of stress. Other times, we can be exposed to stress without being aware of it.

A person under stress secretes hormones that initiate a number of different biochemical processes. One thing that happens is that magnesium is taken out of muscle cells and replaced with calcium.

This causes the muscles to become more rigid, as if to prepare us to defend ourselves against a foe. The danger is that the muscles may lose too much magnesium, triggering the development of cramps or spasms—or even a heart attack. And the magnesium that leaves the muscle cells does not necessarily reenter them once the stress is over. It can be replaced by the consumption of additional magnesium.

Endocrine Gland Disorders

Certain endocrine disorders can contribute to magnesium deficiency. Chief among these is diabetes, a condition in which the body's ability to utilize glucose is severely impaired. Diabetes is a common, devastating disease that throws the body's entire chemistry out of balance. This in turn can cause other serious problems, including overactive thyroid and parathyroid glands and Conn's syndrome, which is characterized by an excessive secretion of the hormone aldosterone and which causes, among other things, the wasting of potassium through the kidneys.

Approximately 90 percent of diabetics are deficient in magnesium. That is why so many of them experience high blood pressure, rapid development of plaque in the arteries, and other cardiovascular problems. Low magnesium levels may also contribute to insulin resistance and carbohydrate intolerance. Many diabetics develop retinopathy, a condition that leads to blindness; many have fatal heart attacks.

Diarrhea

No matter what the cause of diarrhea—whether a chronic illness or a brief encounter with a virus—abnormally high elimination through the intestines can cause significant amounts of magnesium and other minerals to be lost. If you have a case of diarrhea that persists for an unusual length of time, consult with your physician about supplementation to replace the minerals that have been lost. Diarrhea can cause the body to become dehydrated and to lose valuable minerals at the same time.

Diuretic Medications

Surprisingly, according to Dr. W.J. MacLennan, professor of geriatric medicine at City Hospital in Edinburgh, Scotland, about 20 percent of people who take diuretics and are over sixty-five years of age are deficient in sodium.[7] If we know that diuretics can cause the depletion of sodium—much more abundant in our food supply than magnesium—and we know that they cause the excretion of magnesium, we may logically deduce that a major deficiency of magnesium and other minerals also exists among people who take diuretics.

Because of the ability of diuretics to deplete the body of essential minerals, in many cases it is inappropriate to prescribe them to treat high blood pressure without also taking measures to correct mineral deficiencies. In my view, many patients would be better served by eating a nutritional diet low in fats and oils, following a regular exercise program, and using nutritional supplements to correct for deficiencies in the modern food supply.

Alcohol

Dr. Lloyd T. Iseri tells us that one common cause of magnesium deficiency is an abnormally high consumption of alcohol.[8] Alcoholics in general would rather drink alcoholic beverages than eat. When they do eat, their selection of foods is not likely to yield a well-balanced diet. And because of their abnormally high consumption of liquids, they tend to excrete more fluids than nonalcoholics do. Those excretions carry with them a number of vital minerals that are necessary for health.

Even if a person is not an alcoholic, but indulges in an occasional "binge," he or she is in danger of developing a temporary—but momentarily serious—magnesium deficiency. This is what leads to the condition known as "holiday heart."[9] Clearly, moderation is to be preferred in the consumption of alcohol, as in other dietary matters.

Antibiotics and Chemotherapy

In addition to diuretics, other medications can cause a depletion of magnesium. Among these are some antibiotics and chemotherapeutic agents used in cancer therapy. Many cancer patients die of cardiotoxicity caused by the action of chemical agents that are used to kill the cancer cells—that is, they die of heart failure, not cancer. Chemotherapeutic agents are extremely nasty, destructive, toxic chemicals that cause many cancerous cells to die. But they poison many healthy cells also. In addition, they can deplete the body of essential minerals, and we must suspect that at least some of the deaths attributed to cancer occur through the depletion of magnesium and other essential minerals in the heart muscles.

Organ Dysfunction

If your heart, lungs, or liver fails to function properly, you are in danger of becoming deficient in magnesium. Thus, people who are suffering from cardiac arrhythmia, hepatitis, emphysema, or similar afflictions should be aware that they are in danger of developing more serious complications as the severity of their magnesium deficiency increases.

Pregnancy

In many cases, pregnant women develop a form of hypertension (preeclampsia and eclampsia) with other complications as a result of a magnesium deficiency.[10] For this reason the recommended daily allowance of magnesium for pregnant women and nursing mothers is 150 milligrams higher than for other women. Women should be very careful to ingest sufficient amounts of magnesium, as well as other minerals and vitamins, while pregnant or nursing.

Severe Burns

Severe burns can cause the body to lose fluids, and therefore valuable electrolytes. In addition, the pain of a severe burn creates physical and mental stress for the victim. Dr. Iseri tells us that patients with severe (second- and third-degree) and extensive burns can become magnesium deficient.[11] In fact, anyone who is under any kind of agonizing stress, whether physical or psychological in origin, is a candidate for abnormally large losses of magnesium. If left uncorrected, this condition can lead to cardiovascular malfunction.

COULD YOU HAVE A MAGNESIUM DEFICIENCY?

There are a number of signs that point to a magnesium deficiency (medically termed *hypomagnesemia*), but there is no single indicator. If you lack sufficient magnesium, you may feel under tension. Your muscles may be aching or hurting for no apparent reason. You may feel a "heaviness" or pressure in your chest. You may lack energy or feel tired all the time.

Dr. Leo Galland tells us that the most common signs of magnesium deficiency result from a super-excitability of the nerves and muscles.[12] This central condition can cause anxiety; a prickling sensation in the skin; muscular twitching, cramps, or spasms; rapid breathing; heart palpitations; headaches of all magnitudes; and a persistent weakness or debility.

Other signs of magnesium deficiency include high blood pressure, diabetes, low blood sugar, a low concentration of potassium in the blood, irregularities in the heartbeat, and angina. If you get muscle cramps at night, try taking 250 milligrams of magnesium when you get the first signal of the forthcoming cramp. This works for my wife and for others I know who have tried it. Try the same remedy if you suffer from rectal spasms, which are extremely painful.

People with chronic fatigue syndrome usually feel better when they supplement their diets with magnesium.[13] People who have Graves' disease, like former President George Bush and his wife, Barbara, tend to have magnesium deficiencies, too. If you live or work in a stressful environment, or are over fifty years old, you are also a candidate for magnesium deficiency. If you begin to feel better when you take magnesium supplements, you need them. If you have no serious illness, magnesium supplementation can make you feel *great*.

With the techniques available at the present time, the most reliable way to determine your magnesium status is to submit to a metabolic balance study called a magnesium-loading test, which monitors the amount of magnesium consumed in twenty-four hours and compares it with the amount excreted.[14] If the amount of magnesium entering the digestive system closely matches the amount excreted, you can assume that you have a normal level of magnesium in your body as a whole; if you retain a significant fraction of the ingested magnesium, you can assume that your magnesium status is below normal and needs to be elevated.[15] Unfortunately, although the magnesium-loading test is to be preferred, for many people it is too costly and time consuming to be feasible. This leaves several other, less reliable, testing possibilities.

The magnesium content of the red blood cells may be determined. However, there is considerable variation in normal levels from one individual to another, which makes the data difficult to interpret. The use of biopsies to analyze the amount of magnesium in the muscle cells is also possible, but this too has inherent difficulties that can lead to an erroneous interpretation of the data.

Determination of the magnesium level in the blood serum likewise presents problems. Specifically, magnesium can move out of the cells and into blood fluids when an individual is under stress, so a test that determines the amount of the mineral present in serum can give a false impression of its concentration in the muscle cells. For example, during a heart attack, a significant amount of magnesium leaves the heart muscles and enters the bloodstream. This generates a temporary excess in the serum and a deficit in the heart muscles. Over a period of hours, the serum concentration of magnesium returns to normal, and then to a deficiency state. However, the heart's magnesium deficiency remains and can be corrected only through nutritional channels. This is an extreme example, but it illustrates why serum magnesium measurements can be unreliable indicators of the status of magnesium in the muscles, where it matters most.

Clearly, improvements in methods for determining the biological concentrations of magnesium are needed. These improvements must await the development of new techniques that will permit rapid, simple evaluations of the magnesium content in muscle tissues. Fortunately, such procedures are being developed. One of the most promising is the use of nuclear magnetic resonance (NMR), which employs the same kind of technology used in magnetic resonance imaging (MRI). Another that is simpler and less expensive involves the application of new specific-ion electrodes that are sensitive to magnesium.

Improved testing techniques will also have to take into consideration the fact that the concentrations of magnesium in different tissues in the body are not the same, and exchanges of magnesium between them normally take place quite slowly. The extracellular fluids contain only about 2 percent of the body's magnesium (the blood serum contains only 0.3 percent); about 29 pecent of the magnesium in the body resides in muscle tissue; a little more than half is found in the bones.[16] The concentration of magnesium in the cerebrovascular fluid is higher than it is in the blood serum, and experimental data indicate that the concentration of magnesium in the heart's muscles is higher than it is in the skeletal muscles.[17] Thus, to find out the magnesium status of specific tissues, we will need reliable procedures for assaying the magnesium resident in them. Until such analytical methods are ready for routine use, your magnesium status will likely be interpreted from measurements made in the blood serum.

DETERMINING YOUR MAGNESIUM REQUIREMENTS

How much magnesium do you need to consume every day? This is a simple question, but the answer is complicated. It would be wonderful if determining your body's needs were as simple as looking up the recommended daily allowance. However, the RDA is often a feeble guide to the amount of a specific nutrient required for good health. It may be the amount needed to avoid a deficiency disease, but not to support the functions of a really healthy body.

Also, consider this scenario. The RDA values of foods are determined and published by the National Research Council. Industrial interests exert a strong influence in the determination of government policy. Suppose a cereal manufacturer wants to enhance sales by advertising that a particular cereal has 100 percent, or more, of the RDA of several nutrients. It would be less expensive for the company to lobby for low RDA values for those nutrients than to enrich the cereal to meet higher ones. Such interactions between governmental and industrial interests are alleged to be common, and they may be the reason that many RDA values are too low. It's best not to place too much trust in the policies issued by government agencies.

The true required quantity of any nutrient varies from one individual to another and from one time to another in a person's life. Consequently, there are a number of different factors to consider in determining what your specific needs are.

How Much is Enough?

Everyone *should* get at least the recommended daily allowance of each essential nutrient every day. The RDA for magnesium is 350 milligrams for men, 300 milligrams for women, and 450 milligrams for pregnant and lactating women. However, in reviewing magnesium balances in human metabolism, Dr. Mildred Seelig reports that adult men actually need from 6.0 to 6.5 milligrams of magnesium per kilogram of body weight every day.[18] This is the equivalent of about 3.0 milligrams per pound of body weight. Thus, a 150-pound man would need 450 milligrams of magnesium each day, and a man weighing 220 pounds would need about 660 milligrams per day. Men can determine their specific daily requirements for magnesium by multiplying their weight in pounds by 3.0 milligrams. This makes more sense than specifying a single amount for everyone. However, obese people should be careful about increasing their magnesium consumption to very high levels, because magnesium is more active in muscle cells than in fatty tissues. Consequently, they may need less magnesium than their weight would seem to indicate.

Women need somewhat less magnesium than men do.[19] This is related to the fact that women usually have a lower ratio of muscle to fat than men do, and magnesium is more concentrated in muscle cells than it is in fat cells. To have a balanced magnesium metabolism, women need from 5.0 to 5.9 milligrams per kilogram of body weight every day. If we take an average value of 5.5 milligrams per kilogram, we find that most women should need approximately 2.5 milligrams of magnesium per pound of body weight each day. Thus, a woman who weighs 120 pounds needs about 300 milligrams of magnesium per day. Pregnant women and nursing mothers should add 150 milligrams per day to the calculated result.

In addition to being reasonable, these values correspond with the RDA values for average people and compensate for those whose body weights are above or below average. Of course, some people need more magnesium than these numbers indicate. Different people have different requirements. Remember that in some cases, the RDA represents the minimum amount necessary to prevent symptoms of deficiency disease. It does not reflect the amounts of nutrients we need to keep our bodies in top physical condition. Also, it is doubtful that these numbers take into account the low absorption efficiency of magnesium from the digestive tract.

The Absorption Factor

The equation becomes further complicated when you consider that our digestive systems absorb only about 35 to 40 percent of the magnesium we consume. So if we assume that we utilize only 40 percent of the magnesium in our food and supplements, the whole picture changes; in order to actually get the RDA values of 350 milligrams for men and 300 milligrams for women, we would have to consume a total of 875 and 750 milligrams, respectively; pregnant and lactating women would need 1,125 milligrams per day.[20]

Using a 40-percent absorption rate to recalculate Dr. Seelig's values, we find that men should actually consume 7.5 milligrams of magnesium per pound of body weight; women should consume 6.25 milligrams per pound of body weight; and pregnant or nursing women should add about 375 milligrams (instead of the 150 specified in the RDA) to their regular daily total. These amounts may seem to be straining the upper limits of our requirements. However, if one considers that these values apply to the total consumption of magnesium, from both food and supplement sources, they become quite reasonable. (Of course, it is always wise to consult with one or more physicians about nutrient dosages. Above all, make sure that your kidneys are functioning normally before taking any dietary supplements.)

I am now supplementing my diet with 750 to 1,000 milligrams of magnesium per day, depending on my level of activity, and I feel quite well. But you must consider that I am a man who weighs 215 pounds. As a person ages and intestinal absorption decreases, greater quantities of supplements are often needed to maintain good health. Larger people also usually need more nutrients than smaller people.

All of these numbers can be confusing. So the best approach may be to begin with a minimal supplement of magnesium, say 100 to 250 milligrams per day, for a week, followed by a gradual increase, in 100- to 250-milligram increments, until the condition caused by magnesium deficiency is eliminated. Restoration of your normal magnesium status may take several months. That is the way I first started taking magnesium supplements. Because of my serious deficiency, progress in relieving the symptoms was slower than it would have been if I had taken higher doses. However, the slower procedure proved to be safe, and that is what is most important.

In addition, when taking supplements, you must be aware that all of the "milligram" quantities referred to above designate the number of milligrams of magnesium itself, *not* the total weight of any given supplement tablet. For example, the chemical compound magnesium chloride contains only 25.53 percent magnesium, with the rest being chloride ion. Table 2.1 lists the magnesium content of some of the more common compounds that are used for dietary supplementation.

To see what this means in practical terms, look at Figure 2.1, which shows a sample label from a container of dolomite. This label tells us that the total weight of one tablet is 600 milligrams. However, each 600-milligram tablet contains only 78 milligrams of magnesium and 130 milligrams of calcium. That adds up to 208 milligrams of active components. The remaining 392 milligrams are made up of carbonate ion, which has no dietary value. When taking supplements, it is always necessary to determine exactly how much of each specific nutrient they really contain.

Moreover, each individual must regulate dosage according to his or her own specific requirements and situation. For example, a person under stress excretes minerals at a faster rate than he or she would when in a more tranquil mood. I have sometimes found that by midafternoon I am fatigued and have a feeling of tightness

Table 2.1 Magnesium Content of Selected Dietary Supplements

Supplement (Chemical Formula)	Percentage Magnesium	Milligrams of Product Required to Supply 100 Milligrams of Magnesium
Magnesium oxide (MgO)	60.32	166
Magnesium carbonate ($MgCO_3$)	28.83	347
Magnesium chloride ($MgCl_2$)	25.53	392
Dolomite ($CaMg(CO_3)_2$)	13.19	758
Epsom salts ($MgSO_4.7H_2O$)	9.86	1,014
Magnesium gluconate ($Mg(C_6H_{12}O_7)_2$)	5.84	1,712

(or heaviness) in my chest. This condition can be alleviated by taking a nap. If sleeping is not practicable, a tablet containing 100 to 250 milligrams of magnesium invigorates me and relieves the discomfort.

I have also found that I can get the symptoms of magnesium deficiency anytime up to two days after a frustrating event. Now I remedy that by taking a magnesium tablet as soon as possible after such an occasion. This gives me the magnesium I need to compensate for the biochemical reactions that accompany frustrating events. You should experience the same benefit.

Is It Possible to Take Too Much?

The average person's body has a total magnesium content of about twenty-four grams, which translates into somewhere between eight-tenths and nine-tenths of an ounce. If you were to supplement your diet with 800 milligrams (or 0.800 gram) of magnesium per day, you would be taking the equivalent of your total magnesium store in one month's time.

NATURAL DOLOMITE 600 Milligrams	
Each Tablet Contains:	% RDA
Calcium. . . . . 130 milligrams	13
Magnesium. . . . . 78 milligrams	20

Figure 2.1 Sample Dolomite Label

The use of supplements to restore biological concentrations of magnesium is generally quite safe. Dr. Hans-Georg Classen of the Department of Pharmacology and Toxicology of Nutrition at the University of Hohenheim in Stuttgart, Germany, advises us about magnesium's toxicity. He reports that in healthy adult men whose kidneys function normally, there has never been a documented case of fatal poisoning caused by oral consumption of either potassium or magnesium.[21] A fatal dose of magnesium would require between six and twenty-four grams, introduced directly into the intestines.

Let us then establish 1,000 milligrams (one gram) per day as the normal upper limit for regular supplementation of magnesium. In his medical newsletter *Health and Healing*, Dr. Julian Whitaker tells us that most people can tolerate one gram of magnesium per day.[22]

But if you do take too much magnesium, what happens? The first thing you will notice will likely be its action as a laxative. In fact, milk of magnesia, available over the counter in pharmacies, has been used for just that purpose for many years. If you should *seriously* exceed your tolerance for magnesium, your muscles may begin to feel very relaxed, your pulse might weaken, and your blood pressure start to decrease. If you develop any of these symptoms, seek medical help.

Remember, only about 35 to 40 percent of ingested magnesium is absorbed from the digestive tract. Reasonable excesses are eliminated through the kidneys. However, in the unlikely event that you should develop a kidney malfunction or atrioventricular block (an abnormality in nerve signal conduction in the heart), or mistakenly ingest, say, six to ten grams or more of magnesium in a short period of time, call for emergency medical assistance immediately. While you are waiting for help to arrive, write down as much as you can remember of the event that brought on your problem.

I usually take 250 milligrams of magnesium three or four times each day. The regimen of vitamins and minerals I take, among which magnesium is dominant, has restored my vigor. I now feel at least as well as I did twenty years ago—before I became a senior citizen. I move with a quick step and have no muscle stiffness. I can do all of the exercises I did as a young recruit in the United States Army many years ago, and I experience no stiff or aching muscles.

It is distressing to see others my age whose movements are obviously impaired by muscular stiffness, or who have cardiovascular ailments that they simply accept as a consequence of aging. I have learned, however, that almost none of them believe me when I try to advise them of the merits of taking supplements of magnesium and other electrolytes, along with certain vitamins. They are convinced that they must suffer a progressively lower quality of life because they are caught up in the cycle in which all people get old and die. It is true that we die. But if we consume adequate amounts of essential nutrients, get a reasonable amount of exercise and rest, and avoid abusing ourselves with tobacco, alcohol, and rich foods, we can vastly improve the quality of our lives even as we approach the end of our life spans.

TALK TO YOUR DOCTOR

A word of caution should be sounded at this point. I am a chemist—a scientist. I know something about experimental procedures, chemical substances, and how to establish reasonable limits for dosages of nutritional supplements. I have been using myself as the experimental subject in examining the effects of magnesium on the human body, but because of my background, my respect for the potency of chemical substances far exceeds that held by most people.

One should always begin a course of treatment by seeking the advice of a physician. As you know, I did; I struck out on my own only after the prescribed medication proved to be ineffective and inappropriate for my condition. I also had a positive plan supported by biochemical and medical literature and by my own professional expertise. My plan included no medications. I used only nutritional supplements. It is extremely hazardous to expose yourself to unknown medical treatments with no plan or knowledge.

The tests and medications I received while under medical care were standard treatments for my symptoms. Moreover, from the many tests and procedures I underwent, I received valuable information that gave me direction in setting up my own program of supplementation. I have a very high respect for the knowledge, skill, and dedication of the physicians who treated me. However, I believe that patients would benefit immeasurably if physicians would include nutrition as a major consideration in their diagnoses and treatments.

If your physician is reluctant to consider the benefits of magnesium, you may use the information I am giving you to try to convince him or her of its merits. It may be wise to seek several opinions, then decide which physician is right for you. Remember that physicians are hired consultants. True, they have a wealth of knowledge and expertise. But they are also human beings, with all the faults the rest of us have. None of them is correct all of the time. There are, however, some who have superb records. Those are the physicians to seek out and trust.

You may wish to visit a medical library and read the scientific literature available on the subject on your own. Don't assume that your physician has read scientific accounts of magnesium's therapeutic power in the treatment of heart disease and the maintenance of a healthy heart and circulatory system. Medical literature is vast. It is impossible for anyone to read it all, and a physician with an active practice has very little time to spend in libraries.

If you choose to consult some of the references, don't be intimidated by their language. Use a medical dictionary, and go for it. You don't need to understand every word to get the general thrust of an article, but the more you understand, the better.

Magnesium deficiencies can develop gradually over an extended period of time, or they can appear abruptly from some condition such as an alcoholic binge, diarrhea, a severe burn, or a desperate emotional experience. But we can correct those deficiencies by taking magnesium supplements on a regular schedule in recommended amounts. If we fail to heed the warning signs of a deficiency, we can expect to develop more serious ailments—perhaps in a relatively short time. Now let us consider some of the more serious consequences of magnesium deficiency and how replenishing the magnesium in our bodies affects them.

3

A Warning of Heart Disease: Hypertension

Hypertension (high blood pressure) is a disease that afflicts millions of people in the industrialized nations of the world. It occurs when the blood encounters a higher than normal resistance to its flow through the arterial network. In order to understand how high blood pressure develops, and the devastating consequences it can have, we must first understand the normal functioning of the circulatory system.

HOW THE CIRCULATORY SYSTEM WORKS

The cardiovascular system is comprised of the heart and all of the blood vessels in every part of the body. The heart is the muscular organ that propels the blood through the arteries, capillaries, and veins, which form the system of tubes called the vascular network. The heart has four chambers: two on the right side and two on the left.

The right side receives the spent blood as it returns from the veins and sends it to the lungs to expel carbon dioxide and receive a fresh supply of oxygen. The left side of the heart accepts the bright-red, oxygenated blood from the lungs and pumps it into the aorta for distribution through the arteries to all parts of the body. With each beat of the heart, the arteries expand and contract, as the blood pulses through the vascular network to deliver vital nutrients and oxygen to the body's tissues. This happens more than 100,000 times every day in each of us. The arteries carry blood away from the heart, and the veins return it.

The arteries are elastic tubes comprised of three layers of tissue. The inner lining of the arteries is made up of a layer of endothelial (flat) cells, a layer of connective tissue, and a layer of elastic tissue. The middle of the arterial walls is comprised mostly of smooth muscle and elastic tissue intermingled with connective tissue. The outer wall is made of connective tissue and smooth muscle cells.

The pulsing flow of blood from the heart makes the arteries expand. When this happens, the smooth muscles in the arterial walls contract to force the blood toward its destination. This sequence of expansions and contractions is the source of the pulse we can feel at certain pressure points, particularly at the wrist and neck.

As the distance from the heart increases, the arteries subdivide repeatedly and become smaller and smaller until they become microscopic vessels called arterioles. These, in turn, decrease in size until they become capillaries, which are just large enough to allow red blood cells to pass through one at a time. The capillary walls are only one cell thick, which allows oxygen and nutrients to diffuse through them and feed the billions of cells that make up the body. Carbon dioxide and other wastes produced by the cells diffuse across the capillaries' boundaries in the opposite direction and enter the bloodstream. The capillaries are bathed in an intercellular fluid that is the medium of transport between them and the lymphatic system.

Then the capillaries develop into minute, but larger, tubes called venules, which in turn increase in size and evolve into the vessels we know as veins. Although they are thinner than the arteries, the veins have a similar structure. They are collapsible tubes with a sequence of valves, called semilunar pockets, distributed strategically along their inner surfaces. Semilunar pockets prevent the blood in the veins from flowing backward toward the capillaries as the veins return the blood to the heart from all parts of the body.

THE SILENT KILLER

When blood flowing through the arterial network encounters more than the usual level of resistance, the pressure of the blood against the walls of the arteries increases. High blood pressure results directly from a decrease in the elasticity of the arteries, or a reduction in their interior diameters, or both. There are several different conditions that can cause these problems. One is the accumulation of an encrustation, called plaque, on the arterial walls. Plaque builds up from the deposition of platelets, cholesterol, calcium, debris from cells, and other materials that may become entrapped. When this happens, the arteries lose their ability to expand and contract in the normal way, and the heart must work harder to keep the blood circulating. Another possibility is the development of fatty lesions under the lining of the arteries. These and other conditions cause the arteries to become relatively rigid and their internal dimensions to decrease. Spasms in the arteries, although temporary, can also increase arterial rigidity and restrict the blood's passage through the arteries to a dangerous degree. This danger is compounded when an arterial spasm occurs at a site of arterial blockage.

Whatever the cause, a restriction in the flow of blood increases blood pressure and the chance that vital organs may not receive a sufficient supply of life-sustaining oxygen and nutrients. This can lead to heart attack, stroke, and kidney failure, all of which are debilitating and potentially fatal. For this reason, high blood pressure—which by itself is essentially painless—is known as the "silent killer."

We measure blood pressure first by observing the pressure needed to stop the flow of blood during the pumping contractions of the heart. This measurement, the *systolic pressure,* represents the maximum blood pressure attained when the heart contracts and forces blood into the arteries. The *diastolic pressure* is the lowest pressure that occurs in the arteries. It is measured between the pumping pulses of the heart's left ventricle. These two values are reported as systolic pressure/diastolic pressure, for example, 130/85. Many physicians will diagnose a person as having high blood pressure if his or her diastolic pressure is 90 or higher. Consistent readings of over 140 for the systolic pressure are also regarded as abnormally high.

THE DANGERS OF BORDERLINE HYPERTENSION

Recently, Dr. Stevo Julius and his colleagues at the Division of

Hypertension at the University of Michigan sounded a new warning about hypertension. They report that people with even borderline hypertension—those having blood pressures of approximately 130/93—are at risk of developing organ damage.[1] Many of these are people whom physicians will not treat with drugs, but will instead advise to lose weight, stop smoking, and get more exercise. All of that is good, but it may not be enough to effect a cure.

In the Michigan study, the blood pressures of 946 people have been monitored since childhood. Of them, 124 people (13.1 percent) between the ages of eighteen and thirty-eight had blood pressure readings higher than 140/90 and had early signs of heart and blood vessel damage. (Readings of 140 systolic and 90 diastolic are considered the borderline values between high and normal blood pressure.) The average reading in those having borderline hypertension was 130/94. Thus, the major concern seems to be high diastolic pressure. Physicians observed that the hearts of some people in the borderline group pumped less than the average amount of blood with each pulse and did not relax properly between beats. These individuals also had elevated cholesterol, triglyceride, and insulin levels, with high-density lipoprotein ("good cholesterol") readings below normal. One-sixth of them were overweight. These are all danger signs.

CONVENTIONAL TREATMENT

Physicians commonly prescribe diuretics as a treatment for hypertension. Diuretics are medications that decrease the amount of fluid in the tissues by stimulating them to excrete salt and water.[2] This reduces the volume of plasma and the atrial pressure in the heart. As a result, the arteries contain less fluid, and the blood pressure decreases.

Diuretics do reduce blood pressure. However, they also have undesirable side effects. Perhaps their greatest offense is that they can cause the excretion of essential electrolytes at an abnormally high rate.

Electrolytes are the soluble salts dissolved in the body's fluids. These salts, composed of charged atoms and/or molecules called *ions*, are the form in which essential minerals—sodium, potassium, calcium, magnesium, zinc, chloride, phosphate, sulfate, and iodide—exist in the liquids and cells of the body (see Ions and Electrolytes, page 42). When the natural balance of electrolytes is violated, such serious problems as spasms in the coronary arteries, stroke, heart attack, or even sudden death from an abnormal heart rhythm, can result.

Because of their effect on electrolytes, diuretics can in some cases lead to conditions more dangerous than the high blood pressure itself. This is especially true with elderly patients. Dr. W.J. MacLennan tells us that about 20 percent of people over the age of sixty-five take diuretics, and that among the elderly, diuretics are responsible for more adverse side effects than any other prescription drugs.[3]

In view of the emphasis on low-sodium diets for people with high blood pressure, it is surprising to learn that about 20 percent of elderly people who take diuretics for that condition actually suffer from a *deficiency* of sodium. Dr. MacLennan reports that in severe cases this condition can cause people to become weak, confused, partially paralyzed, and unsteady on their feet, and even to lose their mental stability.

From the Medical School of the University of Athens in Greece, Dr. Theodore Mountokalakis warns that as part of their effect on electrolyte balance, diuretics tend to cause excessive excretion of magnesium.[4] This can lead to dangerously low levels of cellular magnesium. As a result, a patient is more likely to develop abnormal heart rhythm or spasms in the coronary arteries. Either of these can be debilitating or fatal.

In addition, each individual diuretic has its own list of undesirable side effects. Thus, the ingestion of diuretics should be closely monitored to make sure the benefits outweigh the risks. The dangers associated with taking diuretics are real, and both patients and physicians should consider them seriously.

WHAT MAGNESIUM CAN DO

As we have seen, the architectural fabric of the arteries contains smooth muscle cells that contract and relax the same way skeletal muscles do. They respond to the same stimuli. The presence of calcium ions is necessary to cause muscles to contract; magnesium ions are essential to induce muscles to relax.[5] If the muscle cells

Ions and Electrolytes

An atom is composed of a dense nucleus—which bears a positive charge—surrounded by electrons that have enough negative charge to exactly neutralize the positive charge of the nucleus. An atom's nuclear charge is the thing that identifies it as a specific element. If the atom loses one or more of its electrons, it assumes a net positive charge and is called a positive ion. Examples are potassium (K^+) and magnesium (Mg^{2+}) ions. On the other hand, if an atom picks up one or more additional electrons, it becomes a negative ion. Examples of these are the chloride (Cl^-) and sulfide (S^{2-}) ions.

When ions dissolve in water, the resulting solution conducts electric current. Because of this, the dissolved ions are called electrolytes. For example, sea water, which contains many different kinds of dissolved ions, will conduct electric current, and the dissolved minerals—composed of ions—are electrolytes. Likewise, the minerals dissolved in our bodies' fluids are also electrolytes.

contain too little magnesium, calcium moves in to take its place, and the muscles develop too much tone, or tension. They contract and tend to be relatively rigid. They lose some of their elasticity. This reduces the interior diameter of the arteries and restricts the flow of blood. It also reduces the amplitude of the arterial contractions that pulse the blood through the vascular network. Even if there is no blockage in the artery, hypertension can result directly from such arterial constriction caused by spasms.

When the infusion of calcium into muscle cells is too extensive, the cells die.[6] This can occur when the magnesium within the cells is depleted to a level that permits calcium to replace it in its normal chemical environment. The replacement of magnesium by calcium within the muscle cells occurs when the body is deficient in magnesium. (In correcting such a condition, however, it appears that we have nothing to fear from taking supplements that contain both calcium and magnesium, because as long as enough of both elements is present, the body distributes or excretes them according to its needs.)

Magnesium tends to control the concentrations and actions of calcium, potassium, and sodium. All of these minerals play a role in the normal contractions and relaxations of the muscles. Thus, hypertension may result from an increase in the concentration of sodium or calcium, or a decrease in the concentration of calcium, potassium, or magnesium.[7]

That blood pressure can be raised by either an increase or a decrease in calcium concentration may appear at first to be contradictory. However, that seems to be exactly what happens.[8] Perhaps this results from differences in hormonal activity. In my case, blood tests showed the calcium concentration in my blood to be too low when I was deficient in magnesium.

The concentrations of potassium and magnesium ions are closely linked. When magnesium concentration decreases, so does that of potassium, and the replenishment of potassium in muscle cells normally can occur only if the magnesium concentration is adequate.[9] Although researchers had previously demonstrated this phenomenon in animal experiments, Dr. M.E. Shils confirmed it in human patients in 1969.[10] That was twenty-five years ago!

An important study done by Drs. Michael Ryan, Robert Whang, and W. Yamalis showed that muscle cells tend to retain magnesium and lose potassium when a deficiency of both elements exists. But the heart's muscles lose potassium at a slower rate than the skeletal muscles do.[11] This and other studies confirm that potassium is more mobile than magnesium, and that all muscles must have a healthful balance of both elements, along with other electrolytes.

Yet many physicians still ignore the role magnesium plays in restoring and maintaining normal levels of potassium in the body's cells. Neither do they recognize the high incidence of magnesium deficiency in the general population, and especially among cardiovascular patients. My own cardiologist—who is very proficient and well respected, and has many patients under his care—failed to recognize that magnesium compounds could relieve my cardiovascular spasms. Because of the interdependence between magnesium and potassium, when a physician prescribes supplementation of potassium without a corresponding dosage of magnesium, we must ask about the possibility of a coexisting deficiency of magnesium, and possibly of other electrolytes. A proper balance of all electrolytes is essential to the healthy operation of all the muscles, including those in the cardiovascular system.

THE IMPACT OF PERSONALITY

Dr. J.G. Henrotte of the Faculté de Pharmacie, Centre Nationale de la Recherche Scientifique, in Paris, France has shown a direct link between so-called type A personality traits and magnesium deficiencies.[12] "Type A" individuals tend to be extroverted and aggressive; "type B" persons tend to be more introverted and relaxed. Dr. Henrotte observed the responses of twenty type A and nineteen type B students to the combined stress of a mental task and a loud noise. Both before and after the stress, he determined the magnesium concentration in their red blood cells, blood plasma, and urine. He also monitored the changes in the calcium content of the participants' plasma and urine; the zinc in their red blood cells and urine; catecholamines (the stress hormones adrenaline and norepinephrine) in their urine; proteins in their blood plasma; and free fatty acids in their blood serum. He found that the average change in these ten variables was 22.6 percent in the type A group and 11 percent in the type B group. In other words, the type A people showed a significantly greater *physical*, as well as psychological, response to stress.

Of the type A participants, 80 percent showed a decrease in the level of magnesium in their red blood cells, 5 percent showed no change, and 15 percent actually had an increase in cellular magnesium. Compared with this, only 44 percent of the type B group had a decrease in the concentration of magnesium in their red blood cells; 56 percent registered an increase. The loss of magnesium from the red blood cells paralleled the loss from muscle tissues and pointed to a possible reason for the higher risk of hypertension among type A persons.

Increased adrenaline secretion in the type A group caused the amount of free fatty acids in their blood serum to increase twice as much as it did in the type B group. The increase in free fatty acids in turn caused a decrease in magnesium, because the fatty acids reacted with magnesium ions to create a new chemical form that could be taken up by the lipoproteins or the fat cells. In this new chemical form, magnesium could not perform its normal functions in the cells.

Through its influence in generating free fatty acids, adrenaline secretion causes a loss of magnesium from the muscle cells. This in turn causes a greater sensitivity to stress. The bottom line is that type A individuals have a greater tendency than type B individuals to be vulnerable to stress and to become deficient in magnesium. This leads directly to the development of high blood pressure and related diseases. Type A people should therefore be especially careful to monitor the concentrations of magnesium, potassium, and calcium in their blood cells.

Another group of people who are especially prone to high blood pressure are those who drink alcohol. Male alcoholics increase their risk of developing high blood pressure with every alcoholic drink they consume. Not surprisingly, there is a direct relationship between the amount of alcohol they drink and the lowering of their magnesium levels. Noticeable elevations in blood pressure usually occur when daily consumption is three to five drinks.[13]

Among women, the story is slightly different. Dr. Luther Clark of the State University of New York Health Science Center in Brooklyn, New York, reports that women who are moderate drinkers have lower blood pressure than either the heavy drinkers or those who abstain. Dr. Clark defines one or two drinks per day as "moderate" and over five per day as "heavy."[14]

Drs. Burton and Bella Altura have shown that excessive consumption of grain alcohol causes a decrease in the concentration of magnesium in the smooth muscles of the circulatory system.[15] There is a consequent increase in calcium, which replaces the depleted magnesium. Thus, the familiar conditions for establishing rigidity in the circulatory system are generated, and the resultant resistance to blood flow causes hypertension.

MAGNESIUM THERAPY WORKS

Dr. Bernard Horn, a surgeon in Benicia, California, has given us one of the most impressive clinical reports describing the use of magnesium to relax the smooth muscles of the vascular system.[16] Dr. Horn reported that he administered magnesium to over 8,000 surgical patients over a period of fifteen years. He gave them three grams (about one-tenth of an ounce) of magnesium sulfate in 1,000 milliliters (slightly more than one quart) of intravenous fluid during a period of six to eight hours. (This three-gram dose of magnesium sulfate contains 590 milligrams of magnesium.) He followed this with an additional dose of one gram of magnesium sulfate (197 milligrams of magnesium) each in subsequent intravenous injections.

Because the magnesium was administered by intravenous injection, it was all absorbed into the bloodstream. If you were taking magnesium orally, you would have to take 1,500 milligrams to get the equivalent of 600 milligrams injected intravenously. This attests to the safety of supplementing our diets with up to 1,000 milligrams of magnesium per day. Remember that the efficiency of digestive absorption is approximately 40 percent (see page 30). If you take 1,000 milligrams of magnesium orally, that means that only about 400 milligrams of magnesium will get into your bloodstream.

Dr. Horn indicated that his procedure had essentially no effect on the blood pressures of patients with normal or low blood pressure. However, patients who had systolic pressures as high as 200 and diastolic pressures as high as 150 usually had normal blood pressures by the time he performed surgery.

Further, Dr. Horn noticed that when he administered the magnesium, the rate of blood flow always seemed to increase throughout the patient's body, regardless of the blood pressure the person had had upon entering the hospital. The increased blood flow was especially evident when he removed veins from the lower limbs. When a vein is removed from a limb, the remaining blood vessels normally contract as the body tries to prevent excessive loss of blood. This response causes constrictions that reduce the flow of blood to inadequate levels and can lead to serious complications. However, in Dr. Horn's experience, the danger of this happening was minimized by the increase in blood flow brought about by the infusion of magnesium.

Numerous experiments (many with animal subjects) support Dr. Horn's results. That he made his observations in the course of working with a large number of *human* patients in traumatic circumstances is highly significant. A study of this magnitude, performed on humans by one individual in a controlled environment, is unusual and valuable. Because he treated more than 8,000 patients, Dr. Horn's observations are quite important—particularly his statement that magnesium always seemed to increase the flow of blood. Let us reverse that statement and ask the obvious question: What happens when the flow of blood is restricted? Depending on the degree of restriction, a person might develop hypertension, cold fingers and toes, a persistent state of fatigue, stage fright, muscle cramps, insomnia, chest pains, congestive heart disease, irregular heartbeats, or even a heart attack or stroke. This speaks strongly for the merits of taking magnesium supplements on a regular basis.

Along with these reports from medical literature, I submit my own experience as described in the introduction to this book. The quality of my life has been greatly enriched since I learned how to maintain an adequate level of magnesium in my system. I no longer require either the diuretic or the calcium-channel blocker I had been taking, and my blood pressure remains in the normal range. It is now four years since my *last* attack of angina. To me, this is positive proof that the use of magnesium and related supplements is a superb way to manage cardiovascular problems.

The information presented here is only a sample of the evidence available, but it makes clear the relationship between dietary magnesium and high blood pressure, which for so many people is the first step on the road to cardiovascular disease. Over time the silent killer can begin to speak up, in the form of a number of serious symptoms and conditions. We will explore some of these in the following chapters.

4

Cardiovascular Disease

Symptoms of heart disease include chest pains, nausea, weakness, restricted physical endurance, a hacking cough, a tendency to faint, shortness of breath, irregular pulse, and high blood pressure. But approximately one-third of all heart attacks occur in people who have never had a symptom of heart disease.[1]

High blood pressure, the silent killer, is usually a sign of diseased arteries—a common consequence of our modern lifestyle. More often than not, a diseased vascular system induces a heart to malfunction by failing to deliver an adequate amount of blood to the heart's muscles. Any condition that decreases the arteries' internal openings or their elasticity will increase the work the heart must do and deprive it of an adequate supply of oxygen and nutrients.

Like all other organs in the body, the heart may become starved for certain nutrients it needs to function properly. For example, the normal conduction of nerve impulses that regulate the heartbeat depends on a proper status and balance of electrolytes. If magnesium and potassium fall below required levels, the rhythm of the heartbeat can be seriously disrupted, and the amount of blood the heart pumps with each stroke can be dangerously reduced.

If the heart itself has no physical defects—like a defective valve or a hole in the wall between two chambers—any disease it develops is likely due to the inadequacy of the vascular network that delivers the oxygen and food supply to its muscles. Now let us look at some of the things that can lead to such a situation.

SPASMS IN THE ARTERIES

When you are exposed to temperatures slightly below your normal comfort level, are your hands sensitive to the cold? Do your fingers hurt or become numb? Do your fingernails turn blue? If so, you may be suffering from Raynaud's phenomenon.

Raynaud's Phenomenon

Also known as Raynaud's disease or Raynaud's syndrome, this is a condition caused by restricted blood flow in the arteries of the fingers and toes.[2] The decreased flow may result from arterial spasms or other causes. A spasm is an involuntary muscular contraction that cannot be voluntarily released. Arterial spasms cause the arteries to contract, and to remain contracted until some involuntary stimulus allows them to relax.

The one-minute cold pressor test is a procedure that can be done to confirm a diagnosis of Raynaud's. It involves immersing the hands in ice water for a period of one minute.[3] In a person with Raynaud's, the cold stimulates the hormones of the nervous system to cause the blood vessels to contract, which increases blood pressure and decreases the flow of blood. Surprisingly, the one-minute cold pressor test can also induce chest pains in susceptible people.[4] This type of chest pain is called Prinzmetal's angina.

How interesting! Some of us can develop chest pains just from allowing our hands to get too cold. This is because, just as Raynaud's phenomenon can be caused by spasms of the arteries in the fingers and toes, Prinzmetal's angina is caused by spasms in the coronary arteries. These arteries contract and severely restrict the flow of blood to the heart. Such an attack should be viewed with serious concern, because it can trigger events that lead to a heart attack. In light of this connection, it seems quite possible that some of the cases in which people suffer heart attacks while shoveling snow might actually be caused by spasms induced by cold fingers and toes, rather than by exertion.

The good news is that such spasms can normally be prevented by maintaining an adequate level of magnesium in the blood and muscles. Drs. L. Cohen and R. Kitzes of Lady Carmel Hospital in Haifa, Israel, applied the one-minute cold pressor test to a group of fifteen patients who had had recurrent attacks of Prinzmetal's angina. Five of them promptly developed anginal attacks, which were relieved within minutes by an intravenous injection of 10 milliliters of 20-percent magnesium sulfate solution (the equivalent of approximately 400 milligrams of magnesium).

Then the same five patients received an intravenous infusion of 96 milligrams of magnesium (in magnesium sulfate solution) over a period of one hour, after which the same cold pressor test was repeated. This time, there was no evidence of angina or of variations in their electrocardiograms. After one more hour of infusion with magnesium sulfate solution, the patients were released.

From these data, we can estimate that approximately 500 milligrams of magnesium ion rendered these patients immune to angina. This, of course, represents the amount of magnesium absorbed into the bloodstream. To achieve the same effect by taking magnesium orally, each patient would have had to take 1,250 milligrams of magnesium, because the efficiency of absorption via the digestive system is only about 40 percent.

Nine of the patients who did not respond to the one-minute test were given a five- or ten-minute cold pressor test. Four of them responded by developing numb, white fingertips. A burning sensation accompanied the rewarming process as their fingertips again turned pink. After these four patients were infused with 96 milligrams of magnesium (in magnesium sulfate solution) over a period of one hour, the doctors repeated the ten-minute cold pressor test. This time, the patients experienced no color change in their fingertips, no numbness, and no burning sensation on being warmed. The infusion was continued for one hour at the same rate, and then the experiment was terminated.

Thus, the spasms responsible for Raynaud's phenomenon seem to be associated with a deficiency of magnesium, even though a clinical deficiency may not be evident from routine tests.

Up to 30 percent of the patients who submit to coronary angiography are shown to be free of coronary artery disease, even though they exhibit common symptoms of heart disease. Coronary angiography is a procedure in which a cardiologist inserts a small catheter into an artery of the arm or groin and threads it up into the coronary arteries; then a dye is injected, and x-ray photographs reveal the degree of arterial blockage. In patients who have all the indications, but no apparent blockage, the symptoms are usually caused by spasms in the coronary arteries, the muscles of the esophagus, or both. All of this is familiar to me because I was diagnosed as having spasms of the coronary arteries, after having been examined by angiography and found to be essentially free of arterial blockage. And according to the cardiologist, my symptoms were impressive.

For as long as I can remember, my fingers and toes have demonstrated the symptoms of Raynaud's phenomenon, but no physician ever told me what caused them to react that way. My symptoms indicate that I have the first, least serious stage of the disease, called the *local syncope*.[5] (At its most serious stage, Raynaud's can lead to gangrene in the affected fingers and toes.) Both my own experience and many accounts written about arterial spasms support the idea that it is quite possible for spasms to occur in a general way throughout the body. People who suffer from even mild symptoms of Raynaud's disease should be aware of this possibility. Fortunately, and quite remarkably, increasing the intake of magnesium to an adequate level eliminates the spasms completely.

I am reminded of a recent visit I made to a periodontist, who performed a surgical procedure after administering copious injections of anesthetic. I was not consciously nervous before, during, or after this experience, but my body reacted as if it had undergone a significant trauma. This became evident on the second day after the event, when I began to chill. Even though the temperature was 70°F and I was wearing multiple layers of warm clothing, I was truly cold. Recalling the Raynaud phenomenon, I took a tablet of magnesium oxide that contained 250 milligrams of magnesium, and within about twenty minutes I was feeling much warmer.

Apparently, either the trauma of the surgery or the anesthetic caused a depletion of magnesium in my muscle cells. Dental anesthetics usually contain compounds that trigger the deactivation of magnesium within the cell structures. They do this for the specific purpose of constricting the blood vessels, so that the anesthetic does not flow away from the site of the injection. Anyone who has problems with dental anesthetics should require the dentist to use an anesthetic that is free of any chemicals that cause the arteries to constrict, such as adrenaline and norepinephrine. Another defensive approach would be to take 250 to 500 milligrams of magnesium immediately before going to the dentist's office.

More recently I had a subtle, but no less real, adverse reaction to a dental anesthetic. This was a lingering effect in which I had trouble thinking clearly. As I was leaving the office, the receptionist asked me for my date of birth, and I had serious trouble trying to get it right. I also had moments of confusion that day and the next. I believe these events resulted from the constriction of blood vessels in my brain, caused by the action of epinephrine and related substances in the anesthetic. If you have ever had a similar problem, be certain to tell your dentist and ask for a compatible anesthetic.

Prinzmetal's Angina

Classic angina pectoris is a syndrome usually characterized by pain in the chest and left arm, and sometimes in the back or the throat, that is initiated by emotional stress or physical activity. Most people with severe atherosclerosis experience angina, as a result of an insufficient flow of blood through the coronary arteries to the heart muscles. Other people, with or without atherosclerosis, may experience a variant form of angina that was described by Dr. Myron Prinzmetal in a research report published in 1959 (hence the term Prinzmetal's angina).[6]

This variant form of cardiac-initiated chest pain differs from its classic counterpart in several ways. The pain is frequently more severe, and its intensity often follows a rhythmic pattern during an attack. Also, neither physical exertion nor emotional excitement—initiators of the chest pain in classic angina—has anything to do with the onset of Prinzmetal's angina. The pain just happens, even if the person is at rest and unexcited. (It may occur while a person is exercising, but the exercise is not the primary cause.) In some individuals, the pain occurs at about the same time each day; in others, it occurs at random intervals.

My anginal attacks generally seemed to take place about two days after I had had a stressful experience. Once I identified that pattern, I was able to avoid the seizures by taking extra magnesium supplements before a magnesium deficiency and resultant spasms had a chance to develop. Now whenever I encounter any exceptional emotional or physiological stress, I increase my daily dosage of magnesium by anywhere from 250 to 500 milligrams, depending on how I feel. This works so well that I now carry magnesium tablets with me at all times.

When a person with significant blockage of one or more of the coronary arteries is also susceptible to attacks of spasms in those arteries, there is a great danger that a spasm may totally cut off the blood supply to a section of the heart. If this condition is not relieved, it can rapidly develop into heart failure.[7] But spasms can cause heart failure to occur even when the arteries are *not* obstructed.[8] Dr. Burton Altura tells us that of people who die suddenly of a heart attack, somewhere between 40 and 60 percent of them have no arterial blockage or history of irregular heartbeats.[9] Two suspected causes of such failure are spasms in the coronary arteries and the occurrence of a severe arrhythmia, such as ventricular fibrillation, either of which can be caused by a deficiency of magnesium.

The most common time for the onset of heart attacks is around 9:00 A.M. Dr. Prinzmetal reported that people who suffered from spasm-induced variant anginal attacks most often experienced them at the same time each day, usually in the morning and late afternoon.[10] *These are exactly the times at which one would expect magnesium deficiency to be greatest.* The morning deficiency is likely a result of overnight fasting and nocturnal transfer of magnesium to the urine. Deficiencies in the afternoon may be caused by depletion of magnesium induced by the stress of the day and not yet replenished by the evening meal. *It is therefore possible that as many as 50 percent of all sudden deaths due to heart failure could be averted by the regularly timed ingestion of magnesium supplements.* In view of this, it seems reasonable that magnesium supplements should be taken at least three times each day: before breakfast, at about 2:00 P.M., and before going to bed.

How Spasms Damage Blood Vessels

In 1981, two teams of researchers reported that spasms in arteries can damage the endothelial cells, which are the cells that form the innermost lining of the arteries.[11] When the lumen (the internal open cross section) of an artery is severely constricted, as it is in a spasm, the endothelial cells that form the lining of the arteries are compressed to the extent that some are dislocated, some are distorted, and some even merge with other cells so that their contents become intermingled. When the spasm ends and the artery relaxes, the cells are thus left in a state of disarray.

The presence of damaged sites in the endothelial layer sets the body's repair mechanism in motion. Blood platelets and specialized white blood cells cover the afflicted areas, and fibrinogen and other proteins, in addition to low-density lipoproteins or lipoprotein(a), become involved in the formation of a "patch" to protect the inner cells while the healing processes restore the lining to its normal condition.[12]

In many cases, this repair patch on the inner wall of the artery extends slightly into the open space within the artery and slowly grows to close off a significant portion of it. This barrier, called a *thrombus*, may become large enough to restrict the flow of blood. A portion of it may also break off and move through the bloodstream to a narrow spot in an artery, where it can block the blood's flow. Such a mobile particle is called an *embolus*.

At the University of Massachusetts Medical Center, Drs. Isabelle Joris and Guido Majno induced spasms in mouse arteries by treating exposed arteries with the catecholamine hormone norepinephrine. The spasms occurred quickly and damaged the arteries in the manner just described.[13] At the Hebrew University in Jerusalem, Israel, Dr. S. David Gertz and his colleagues achieved similar results by forming arterial constrictions. They tied suture thread around the arteries of mice and drew the thread up until the internal diameter of the arteries was reduced by about 50 percent.[14] By use of scanning electron microscopy, both groups observed the same type and extent of damage.

In later work, Dr. Gertz and his coworkers found that magnesium, administered intravenously before the arteries were constricted, caused platelet deposition to be limited to one or two discontinuous layers.[15] No thrombi formed. In contrast, microthrombi were observed in the arteries not treated with magnesium.

These results are a clear demonstration of the antiplatelet activity of magnesium. Much has been written about the inhibition of platelet deposition by other agents, especially aspirin, but this property of magnesium has been largely ignored. It is a fertile area for research. Dr. Gertz's findings suggest that people can avoid taking aspirin, with its irritating side effects, if they simply maintain an adequate level of magnesium in their systems. And magnesium is an essential component in our biochemical makeup; aspirin is not.

Treatment for Arterial Spasms

Nitroglycerin is a vasodilator, a drug that dilates the blood vessels and increases the flow of blood. Both classic and variant angina respond to nitroglycerin, but it often takes fifteen to twenty minutes to relieve the pain and can leave a person weaker and more exhausted than usual for the next twenty-four hours or so. Nitroglycerin's beneficial effect is short-lived, and the drug may need to be administered in some sort of timed-release form. Also, an annoying headache accompanies any relief it gives.

Dr. Prinzmetal found that a drug called Nylidrin, also a vasodilator, was effective in relieving the pain from variant angina. Since this drug was introduced back in 1953, it has been replaced by newer drugs called calcium-channel blockers and beta-blockers.

An abnormal infusion of calcium into any kind of muscle cells causes them to develop tone, in the form of a partial, persistent contraction. When this happens in the vascular network, especially in the coronary arteries that supply the heart muscles, the operation of the heart can be seriously impaired. Two German researchers, Drs. Erwin Neher and Bert Sakmann, received a Nobel prize in 1991 for establishing beyond any doubt that mineral ions enter the cells by way of certain passageways, called ion channels, that penetrate the cell walls.[16] Ion channels are the tiny openings through which cells transfer chemical species across their membranes. The ion channels important to the present discussion are the ones known as calcium ion channels, or simply calcium channels. The calcium channels seem to be guarded by chemical species containing magnesium, which act as natural chemical "valves," or calcium-channel blockers.

This discovery has extremely broad applications in all sciences related to biological and biochemical processes. As we might expect, pharmaceutical companies have made use of this knowledge as a basis for developing new drugs—synthetic calcium-channel blockers—that limit the migration of calcium into the cells and inhibit the formation of excessive concentrations of calcium inside the cells.[17] And physicians today routinely prescribe these drugs to prevent the abnormal accumulation of calcium in muscle cells. Examples include Calan, Isoptin, Procardia, Adalat, and Cardizem, the drug I took. (Of course, new drugs are being developed all the time.) All of these drugs operate in some way to inhibit the passage of calcium from the extracellular fluid through the cell walls and into the cells. By means of this action, calcium-channel blockers relax the smooth muscles of the blood vessels, increasing the flow of blood and reducing the amount of energy needed to maintain a healthy level of circulation. Reports indicate that they also decrease, or eliminate, the incidence of spasms in the arteries. They tend to reduce high blood pressure and slow the speed with which electrical signals are transmitted in the heart as well. These properties should combine to decrease the severity and frequency of anginal attacks.

The bad news is that the synthetic calcium-channel blockers can have severe side effects. Cardizem can cause liver damage, abnormal heart rhythm, temporary memory loss, and even temporary stoppage of the heart. It can also harm a developing fetus. Possible side effects of Calan and Isoptin include liver damage, abnormal heart rhythm, and heart failure. Adalat and Procardia can produce palpitations, nervousness, breathing difficulty, and muscle cramps. And all of these drugs can cause nausea, fatigue, headache, dizziness, and low blood pressure.[18]

Another type of medication prescribed for many heart problems is the beta-blocker. Beta-blockers are agents that block the chemical sites in the arteries and heart that interact with the stimulating hormones secreted by the adrenal glands. Thus, they slow the heart and make it less susceptible to stimulation caused by emotional excitement. In this way they reduce the heart's workload. Physicians also prescribe these drugs to control blood pressure and even to treat glaucoma. The generic or chemical names of beta-blockers usually end in *olol*. Examples include metoprolol (marketed as Lopressor), nadolol (Corgard), pindolol (Visken), propranolol hydrochloride (Inderal), and timolol maleate (Blocadren).

Beta-blockers too have undesirable side effects, including decreased heart rate, depression, bronchial spasms, and aggravation of congestive heart disease. There are others, but these should suffice to stimulate you to question both your physician and pharmacist closely before taking a beta-blocker.

The side effects caused by calcium-channel blockers and beta-blockers make magnesium a much more attractive treatment for angina. As we have noted, magnesium plays a major role in inhibiting platelet aggregation, relaxing arterial smooth muscles, relaxing skeletal muscles, normalizing blood pressure, increasing circulation, preventing arterial spasms, reducing the heart's workload, and relieving angina. Magnesium is the body's natural calcium-channel blocker. And under normal conditions, with reasonable dosage, there should be *no* side effects. Of course, you should always consult with your physician, especially if you have a problem such as impaired kidney function or atrioventricular block (a condition that involves a deficiency in the electrical conduction between the heart's atrial and ventricular chambers), or if you are taking medications.

In view of magnesium's ability to cause muscles to relax, it seems prudent that persons susceptible to arterial spasms, including attacks of Prinzmetal's angina, should take magnesium supplements to achieve higher levels of magnesium concentration. This simple, inexpensive routine, properly monitored and administered, should eliminate the problem. It worked for me.

Spasms Resulting from Physical Defects

Up to this point I have been talking about people who experience arterial spasms caused by a deficiency of electrolytes, especially magnesium. I have been assuming that they were otherwise normal in all respects, with the possible exception that some might have atherosclerosis. However, I must point out that in some cases, spasms in the coronary arteries can arise from one of a variety of structural abnormalities. These include defective heart valves and various connective tissue disorders. Such conditions are *not* the subject of this discussion, and would not be improved by the ingestion of additional magnesium. This is why, if you are having spasms, you should consult with one or more physicians, have any necessary tests performed, and determine the nature and source of any ailment. Only then can intelligent decisions be made regarding treatment.

ABNORMAL HEART RHYTHM

Have you ever noticed your heart beating in an irregular pattern for several seconds? Perhaps you had overeaten, or drunk too much coffee or another beverage containing caffeine. Maybe you were suddenly frightened. You might have drunk one beer or Manhattan too many. Maybe—for no apparent reason—your heart just began beating faster than normal for your level of activity. On the other hand, you may have a chronically abnormal heart rhythm.

Abnormal heart rhythm—or a troublesome and potentially dangerous cardiac arrhythmia—is a condition in which the nerve signals that control the rhythm of the heartbeat are transmitted in unusual patterns. They may or may not be coordinated with the normal pulses necessary to pump blood effectively throughout the body. Some arrhythmias are merely annoying, and point to no serious health problems. Others are quite serious—even life threatening. With some, the irregular beat can be felt; with others, it cannot. Some arrhythmias produce symptoms such as nausea and fatigue, rather than any feeling in the chest; some may produce no discernible symptoms at all.

If you are at rest and your heart is beating faster than normal (giving you a pulse of over 100 beats per minute), you are experiencing something that is commonly called *tachycardia*. If your heart beats with a noticeable increase in force, whether or not the rhythm is irregular, you have *palpitations*. *Fibrillation* is a condition that occurs when the nerve impulses regulating the heart convert its normal rhythmic beating into a rapid twitching. With a severe attack of some types of fibrillation, no useful amount of blood cycles through the circulatory system, and death is certain if normal rhythm does not resume in a very short time.

Dr. Robert Whang of the University of Oklahoma Health Science Center and Veterans Medical Center in Oklahoma City reports that deficiencies of either potassium or magnesium ions, or both, can lead to cardiac arrhythmia.[19] And if the concentration of magnesium in the muscle cells is below normal, the restoration of normal levels of potassium is severely impaired.[20]

Arrhythmias that can be treated with magnesium include rapid heartbeat, premature beats, multifocal atrial tachycardia (in which the nerve signals that cause one or both atria to contract fire off from several different locations), and ventricular fibrillation.[21] Those that respond to potassium include ventricular tachycardia, ventricular fibrillation, and ventricular ectopic beats.[22] Ectopic beats occur when the locations at which the signals for the heartbeats originate are displaced from their normal positions.

Magnesium and potassium together form a winning combination, with magnesium as the dominant element. About 40 percent of patients who are deficient in potassium are also deficient in magnesium.[23] Therefore, initial treatment for arrhythmia should include supplementation with both magnesium and potassium, not just potassium, as is common practice. *The body's normal potassium levels usually cannot be restored as long as magnesium levels remain subnormal.*

The people who seem to be at greatest risk of developing arrhythmia are alcoholics, people who are taking diuretics or digitalis (or both), and those who are suffering from congestive heart disease. It is sad to realize that diuretics and digitalis are routinely prescribed for heart patients, without any thought of taking measures to correct the electrolyte depletions they cause.

Dr. Lloyd T. Iseri of the University of California Medical Center in Orange, California, documented four cases in which patients who had not been helped by medications responded to magnesium sulfate (Epsom salts) injections. These patients suffered from such afflictions as acute alcoholism, hypothyroidism, coronary artery disease, congestive heart disease, coronary artery spasms, and overdose of digoxin (a form of digitalis). All but one had either fibrillation or some other arrhythmia so severe that they received repeated electric shock treatments or "fist pacing" to restore rhythm. It would be hard to imagine people with more hopeless conditions. However, in each instance, recovery occurred after intravenous injections of magnesium sulfate were administered.[24]

SUDDEN CARDIAC DEATH

Even people who appear to be perfectly healthy can have concealed nutritional deficiencies. Recall that I had my last and most severe attack of angina when I was feeling and looking quite well. Fortunately, I recovered. Some folks are not as lucky as I was.

Have you ever known anyone who appeared to be in excellent health, but who died suddenly of a heart attack? I had two good friends who exercised regularly and rigorously, only to die suddenly of heart failure—one while he was out jogging. People who engage in vigorous exercise—even if they believe they are in good health—should always be alert to the possibility that the stress of strenuous exercise can place them in a temporary state of magnesium depletion, with potentially devastating results. Other types of stress can have the same effect. People who experience great stress—whether voluntarily, as with exercise, or involuntarily, from external sources—should be especially careful about maintaining adequate levels of magnesium.

CONGESTIVE HEART DISEASE

In addition to suffering from vascular diseases, some people are also confronted with abnormal heart function. Among the most serious of these disorders is an abnormal rhythm of the heartbeat. Some people suffer from enlarged hearts; others have a condition called left ventricular hypertrophy, in which the walls of the heart's left ventricle thicken as a result of the heart pumping against the increased resistance of high blood pressure. The pericardium (the outer covering of the heart) can become infected. The list goes on and on. But let us now focus on a common condition that can take you progressively from good health to a wheelchair, and then to a coffin: congestive heart disease.

When the heart's ability to pump is seriously impaired, fluid accumulates in the ankles, lungs, and other tissues. This condition is called congestive heart disease. Although the disease is named for its outward symptoms, its fundamental cause is weakness of the heart, which may arise from a number of different causes. This condition is also a common cause of cardiac arrhythmia.[25]

In congestive heart disease, the output of blood from the left ventricle, which sends the blood out into the arteries to circulate throughout the body, is severely reduced. Coexisting with this condition is a thickening of the walls of the heart and a tendency to form fibrous tissue. The net result is that the heart uses energy very inefficiently, placing it in a serious state of work overload.

In order to maintain adequate blood pressure despite the heart's reduced output, the vascular system responds by developing tone, or tension—a condition akin to chronic arterial spasm. But this further increases the work the heart must do in order to pump the required amount of blood. As a result, the person experiences edema (accumulation of fluid in the tissues), shortness of breath, muscular weakness, and persistent fatigue. Commonly coupled with this is heart arrhythmia, which frequently leads to sudden death. The situation is grim at best.

Congestive heart disease is commonly treated by the administration of diuretics to reduce the volume of fluid in the body, as well as other drugs, such as beta-blockers, that are supposed to maintain a normal heart rhythm. This, of course, treats the symptoms and not the underlying cause of the disease. In fact, *the use of these medications tends to accelerate the patient's decline in health.*[26]

One commonly prescribed diuretic, Lasix, is a very strong medication designed to cause the elimination of fluids. Unfortunately, it also eliminates essential electrolytes, among which are magnesium and potassium. It has other potential side effects, too, including persistent fatigue, thirst, drowsiness, and muscle cramps.[27] The cramps—or spasms—along with most of the other side effects, can result directly from the depletion of electrolytes. This is a powerful drug that necessitates careful supervision as well as supplemental replacement of the vitamins and minerals. Beta-blockers can cause a decreased heart rate, depression, bronchial spasms, and aggravation of congestive heart diesase.[28] If your physician prescribes any of these drugs for you, inquire about the need for supplemental minerals and vitamins.

Physicians report that patients with chronic congestive heart disease have serious deficits of potassium, magnesium, sodium, and other electrolytes, which diuretics can only make worse.[29] The imbalance and deficits in electrolytes are further exacerbated by the negative effects of abnormal secretions of hormones. For example, an increased level of adrenaline circulating in the bloodstream depresses the levels of potassium and magnesium and serves to make any existing electrolyte deficiencies worse.[30] And since the behaviors of magnesium and potassium are interrelated, a deficit in one implies a probable deficit in the other. Recall that both of these minerals are necessary for the maintenance of a normal heart rhythm.

Fortunately, some cardiologists have become aware of the possibility that the incidence of sudden death among patients with chronic congestive heart disease may be minimized if their electrolyte levels are brought into balance. Among these enlightened individuals are Drs. Milton Packer, Stephen Gottlieb, Paul Kessler, Wai Lee, and Mark Blum of the Mount Sinai School of Medicine of the City University of New York, and Drs. Burton and Bella Altura of the Downstate Medical Center in Brooklyn, New York.[31]

When conventional medications proved totally ineffective, another enlightened physician, Dr. Lloyd Iseri, successfully treated a patient with congestive heart disease by intravenous infusion of magnesium sulfate.[32] This patient was in the hospital and had such severe arrhythmia that his heart required periodic pacing. The magnesium restored normal heart rhythm.

An essential chemical substance in all muscle is a compound formed by the chemical combination of magnesium with adenosine triphosphate (ATP). This compound (Mg:ATP) is necessary to maintain cellular health. In studies of decreased blood flow in the liver, Dr. I.H. Chaudry at Michigan State University, and his colleagues at Johns Hopkins University and St. Louis University Medical Center, found that when they infused magnesium and ATP separately into the blood, Mg:ATP did not form in the cells. However, all three components and normal cell function were restored when they administered the compound Mg:ATP intravenously.[33] In another research report, Dr. Chaudry and others tell us that *infusion of Mg:ATP intravenously into normal human volunteers increased cardiac output by as much as 131 percent, without increasing their blood pressure.*[34] This should therefore be an effective procedure for minimizing damage to the heart muscles during and soon after a heart attack. It should also be an effective treatment for patients with congestive heart disease.

Again, a select group of dedicated scientists, who are also physicians, recognize the essential nature of magnesium. One can only guess how many lives might be saved if all physicians were to adopt this knowledge and expertise as part of general medical practice. And keep in mind that this is not a matter of administering medications; it is a matter of restoring normal cellular components and advising patients about proper nutrition.

Sadly, most people with congestive heart disease are subjected to the conventional medical treatment, without any attention to nutrition, and many go on to develop other complications, including adult-onset diabetes. These people need to know the importance of abandoning the typical fat-laden American diet and of taking supplemental vitamins and minerals. Otherwise, they will very likely—slowly but surely, and under the care of a physician—decline, and before long take up residence in the family's burial plot. Of course, I cannot say that any particular patient will survive the ravages of congestive heart disease if he or she takes a prescribed set and amount of nutritional supplements, but we do know that such a program will give an individual a much better chance of recovery.

Patients with congestive heart disease who are confined to a hospital bed and are being fed through an intravenous tube are totally at the mercy of their physicians and nurses. If the intravenous solutions contain an inadequate supply of the vitamins and minerals needed on a daily basis, the patient in effect slowly starves to death. Drs. Milton Packer, Stephen Gottlieb, and Paul Kessler tell us that the combination of neurohormones circulating in the bloodstream and the presence of an electrolyte deficiency (especially deficits of potassium and magnesium) make congestive heart disease worse.[35] They also tell us that the drugs used to control arrhythmias are of little help and may even harm patients.

Congestive heart disease is a severe condition that, under present conventional treatment, offers little prospect of recovery. Many of those who are so afflicted die suddenly, supposedly from some form of cardiac arrhythmia. Dr. Packer and his coworkers tell us that congestive heart disease has a greater potential for generating arrhythmia than any other disorder in cardiovascular medicine.[36] Of course, it is also possible that spasms in the coronary arteries play a role in many of those deaths.

Arterial spasms, cardiac arrhythmias, and congestive heart disease are three different ailments that can lead to the same result: heart failure. Each of these disorders can be caused by a deficiency of magnesium and other essential nutrients. And in many cases, they can be cured by correcting these nutritional deficiencies. Now let us turn to a discussion of what physicians call the "end points" of heart disease—heart attack and stroke.

5

Heart Attack and Stroke

In the United States alone, approximately 1.5 million people suffer heart attacks each year. Of those, about 500,000 die—nearly two-thirds of them before they reach a hospital. Those who have survived a heart attack or have angina pectoris number about 6.2 million. Strokes occur less frequently; about 145,000 Americans die of strokes each year, and approximately 3 million living today have survived one or more strokes.

WHAT IS A HEART ATTACK?

We all know and use the term, but what exactly happens when a person has a heart attack? A heart attack is a general term for a seizure in which the heart's functioning is abruptly and seriously impaired. Symptoms that can accompany a heart attack include a severe crushing feeling in the chest, nausea, blurred vision, profuse perspiration, weakness, and fainting.

Like all other human tissue, the heart's muscles are fed by the flow of blood through a system of arteries, capillaries, and veins. When the supply of blood reaching the heart's muscles falls below a certain level, the heart muscles become starved for blood, and the oxygen and nutrients it carries. This is a general condition known as *ischemia* (pronounced iss-kee'-me-a).

When the ischemic condition becomes severe enough, heart tissue dies. In medical terms, the dead tissue is described as being *infarcted*. An infarction is a macroscopic (visible to the naked eye) area of dead tissue. Thus, a doctor may refer to a heart attack as a myocardial (heart muscle) infarction, or MI. It is important to realize that, unlike congestive heart disease, a heart attack is not related to the amount of blood the heart is pumping. Rather, it is related to the amount of blood the heart tissues receive from the circulatory system.

The heart muscles may become starved for blood as a result of a number of different mechanisms. Three of the most common ones are:

1. The arteries feeding the heart may become blocked or narrowed, preventing enough blood from passing through them to get to the heart. This can happen in a number of different ways. Physical blockages can be caused by fatty deposits under the lining of an artery; deposits of plaque on the inside of an artery; the formation of blood clots in an artery at points where the blood flow is restricted; or a combination of any or all of the three. Blockage or narrowing can also result from spasms in the arteries, in which the arteries contract and become tense, restricting the flow of blood. A spasm that occurs at the site of a blockage in an artery can totally shut off the flow of blood to the heart and lead to a fatal heart attack.
2. An arrhythmia (irregular heartbeat) may develop, causing the heart to pump only a fraction of the blood needed for it to receive an adequate supply. One severe type of arrhythmia, called ventricular fibrillation, causes a rapid, irregular twitching of the heart muscles that renders the heart completely ineffective at pumping blood. If this condition lasts longer than a few minutes, death is certain.
3. An aneurysm (a weak section in a blood vessel) may rupture, causing a massive hemorrhage. When this happens, the circulatory system ceases to function.

The relationship between arrhythmias and magnesium was discussed in the previous chapter. We will explore here the first of these scenarios and, to a lesser extent, the third.

SETTING THE STAGE FOR A HEART ATTACK

Heart attacks don't just happen. The conditions that lead to a heart attack usually take years to develop, and we are now able to recognize some of the things that put people at risk. Perhaps the major identifiable risk factor is an unfavorable (that is, relatively high) ratio of a certain group of molecules in the blood, called low-density lipoproteins, to another group, called high-density lipoproteins, especially when this ratio exists together with an inadequate level of vitamin E and other substances known as antioxidants. It is the low-density lipoproteins that appear to be the source of active cholesterol.

Cholesterol

Contrary to common perception, cholesterol is *not* a fat. It is actually a chemical compound that is classified as a steroid. Cholesterol is the most abundant steroid in the body and is essential in maintaining healthy cells. Because it is soluble in fats, however—and therefore must be transported by fat molecules in the bloodstream—it is also classified as a lipid. The lipids are a group of biochemical substances that are soluble in the same solvents that commonly dissolve animal fats and vegetable oils. This is why cholesterol is associated with fats in the human body.

Cholesterol moves in the bloodstream throughout the body for distribution to the individual cells, but because it is insoluble in water it cannot dissolve in the blood plasma. Another complicating factor is that cholesterol is a solid substance that melts only at a temperature higher than the boiling point of water. Consequently, cholesterol must migrate in the bloodstream by means of some type of carrier mechanism. The body solves this problem by having cholesterol react with fatty acids to form liquid compounds called *cholesteryl esters*. These substances coalesce into tiny oily droplets that become intimately associated with the complex structure of one or more very specific protein molecules. The surfaces of these protein molecules are compatible with water, so by means of this association the cholesteryl esters can travel in the bloodstream. The tiny particles that result are called *lipoproteins*.

If a lipoprotein molecule is comparatively large, its density is relatively low (even though it is still more dense than water), and it is therefore designated a *low-density lipoprotein* or LDL. When the protein occurs as a relatively small molecule, its density is higher, and it is known as *high-density lipoprotein* or HDL.

LDLs are heavily laden with cholesterol, which they carry to all parts of the body for utilization in biochemical processes. HDLs, in contrast, carry a relatively small load of cholesterol. They scavenge excess cholesterol from the blood and carry it back to the liver. When these molecules and all the related biochemical processes are properly synchronized, the cholesterol seeks out the destinations where it is needed to perform its normal functions, and excess cholesterol attaches itself to HDL molecules and goes back to the liver, where it is stored.

Arterial Plaque

As long as these processes maintain a state of balance and the LDL does not react chemically with oxygen (or anything that behaves like oxygen) in the course of its travels, no excess cholesterol will deposit in the blood vessels. But physicians tell us that the danger of cholesterol being deposited in the arteries increases as the percentage of LDL increases and that of HDL decreases. When the amount of low-density lipoproteins is too high and there is a deficiency of high-density lipoprotein, there is not enough HDL to scavenge the excess cholesterol, and it tends to be deposited in the blood vessels, where it contributes to the formation of plaque. Not only are cholesterol and cholesteryl esters deposited, but entire LDL molecules become part of the plaque's structure.

Some of the major components in the bloodstream that contribute to the building up of plaque are large white blood corpuscles, cholesterol, LDL, platelets deposited from the blood, fibrinogen (a blood protein necessary for clotting), and calcium. Plaque deposits gradually grow and often become dangerous obstructions in the arteries. Sometimes the deposits become so rigid that they cause cracks to form in the arteries; these in turn are filled with more platelets, fibrinogen, calcium, etc., as the body attempts to repair the damage.

Blood Clots

Another heart-attack scenario involves the adhesiveness of the platelets (flat immature cells that help form clots) and the turbulent flow around a thrombus, which is a localized "bump" on an artery's wall that may result from the body's attempts to repair an injury there. Increased activity develops around the thrombus in the bloodstream—much as a river passing over rocks becomes turbulent—and this seems to increase the tendency of platelets to adhere to the arterial walls, to the thrombus, and to themselves, forming a clot. The additional obstruction caused by a blood clot can restrict the flow of blood to dangerous levels and cause sufficient oxygen deprivation to precipitate a heart attack.

The "stickiness" of the platelets is a major factor in the formation of clots. For this reason such anticoagulants as aspirin and warfarin (a prescription drug sold under the names Coumadin, Carfin, Panwarfin, and Sofarin), which keep the platelets from sticking together, are routinely prescribed for patients with heart disease. Warfarin is a powerful drug whose dosage must be very carefully regulated. Because of its anticlotting properties, it also tends to cause internal bleeding.

PREVENTING A HEART ATTACK

The only sure way to have completely unobstructed blood vessels is to have the right genes. Unfortunately, this is something you can't control. The next best tactic is to monitor your diet carefully and take the right nutritional supplements in the correct amounts. Eating a diet low in cholesterol and saturated fats is one of the best preventive approaches we know of at this time. But, of course, this requires a discipline that many people do not have. There are some cholesterol-lowering drugs on the market, but they are available only by prescription, they are expensive, and they have very undesirable side effects.

But don't despair. The situation may not be as bleak as it seems.

The Importance of Antioxidants

Recent work has cast new light on the process of cholesterol deposition. We now know that all cholesterol is not created equal. Only cholesterol that has reacted chemically with oxygen or a similar kind of chemical is readily deposited in the arteries. This is true even of cholesterol from LDL, the so-called "bad cholesterol." And there are natural agents known as antioxidants—including vitamin A, vitamin C, vitamin E, and beta-carotene—that tend to inhibit the oxidation process. *This is important because it seems that one can find protection from the deposition of cholesterol by maintaining adequate levels of antioxidants in the bloodstream.*[1] Dr. Daniel Steinberg and his coauthors explain this in an excellent review article in *The New England Journal of Medicine.*[2]

The late Linus Pauling, Ph.D., twice a Nobel laureate, was for years a strong advocate of the curative powers of vitamin C. The medical community, however, largely rejected his claims. I am convinced that this has been a big mistake. Dr. Pauling was one of the greatest scientists of our day. He was born in 1901 and was active in scientific research until his death at the age of 93! I know of no medical researcher who can match the lifelong accomplishments of this outstanding chemist. It was he, for example, who discovered the structure of protein, a discovery that led to one of his Nobel prizes.

Dr. Matthias Rath and Dr. Pauling proposed that the real culprit in the formation of arterial plaque is *not* low-density lipoprotein, as some researchers theorize.[3] They said that they had identified a different lipoprotein, called lipoprotein(a) (abbreviated Lp[a]), as the agent that becomes attached to the arterial walls and forms plaque.

Lipoprotein(a) was first reported by Dr. Kåre Berg in 1963.[4] It forms in the liver from the combination of a substance called apoprotein(a) and LDL. In their report on their findings, Drs. Rath and Pauling tell us that in order for plaque to form, the apoprotein(a) must be chemically attached to LDL. They also point out that vitamin C inhibits the formation of apoprotein(a). Therefore, they reason, if apoprotein(a) cannot form, neither can lipoprotein(a), nor can the plaque that results from it. Accordingly, Dr. Pauling recommended that everyone take massive doses of vitamin C on a daily basis. (I myself am inclined to believe that his own practice of taking a large quantity of vitamin C every day may have had something to do with his longevity.)

Several discrepancies exist between the report by Drs. Rath and

Pauling and the findings of Dr. Steinberg and his colleagues. Specifically, Dr. Steinberg believes that vitamin E is dominant as the protective agent; Drs. Rath and Pauling give full credit to vitamin C for keeping the arteries free of plaque. Both, however, recommend the use of antioxidants. It makes the most sense, therefore, to supplement our diets with *all* of the antioxidants we have been considering here.

Vitamins E and A and beta-carotene are all oil soluble. As a result, they are carried in the bloodstream mainly in the lipoproteins, as is cholesterol. Vitamin C, on the other hand, is a water-soluble organic compound and is present mostly in the watery part of the blood. Because of the different chemical natures of these vitamins, vitamin C can regenerate at least some of the vitamin E that has reacted with oxidized LDL and allow it to act once again as an antioxidant.

It has been observed that chewable vitamin C tablets apparently have the ability to chelate (bind with and remove) calcium, and therefore erode tooth enamel. In view of this, it certainly seems possible that vitamin C might be involved in removing calcium from plaque in the arteries. This is a chemical property of the vitamin C molecules themselves, so vitamin C should tend to extract even some of the calcium that is deeply imbedded in the plaque. This could then soften the deposit and open its structure up to the action of enzymes that could further destroy it.

To gain maximum protection against cholesterol and plaque, therefore, one should maintain an adequate level of vitamin E, vitamin C, and beta-carotene in the blood at all times. The body uses beta-carotene to make vitamin A, so as long as beta-carotene concentrations are adequate, it is not necessary to worry about vitamin A.

Diet and Exercise

Many people should be able to improve their chances of avoiding heart disease by taking supplemental beta-carotene and vitamins A, C, and E, in addition to magnesium. Even if you choose not to make any drastic alterations in your eating habits, the addition of these supplements should make a dramatic difference. However, if you wish to maintain truly optimal health, you should consider changing your diet to one that is high in complex carbohydrates and low in both protein and fat. Fats are particularly harmful because they have more than twice the caloric content of either protein or carbohydrates. Fats contain nine calories per gram; one gram of either protein or carbohydrate, on the other hand, contains only four calories. If your diet consists of 40 percent fats or more, you are likely to be quite a bit heavier than you should be, a condition that, in addition to predisposing you to heart disease, can lead to many other different health problems.

Following a macrobiotic diet, or a diet like the one recommended by Nathan Pritikin, Dr. Julian Whitaker, and Dr. Dean Ornish, offers protection from cholesterol deposition.[5] However, maximum protection should be achieved by means of both dietary restraint *and* judicious use of dietary supplements, as well as exercise. This is the approach taken by Dr. Whitaker.

Finally, much of what we read these days concerning health would lead us to believe that aerobic exercise will prevent a heart attack. A healthy exercise regimen is certainly a good idea, provided it is within your capabilities. But you should always be alert to the possibility that the stress of overexercising can place you in a state of temporary magnesium depletion, which creates an increased risk of an attack of coronary arterial spasms or perhaps arrhythmia. Similarly, all people who are under extreme stress, and thus at a greater risk of magnesium deficiency, should seriously consider taking corrective action. The simplest way to do this is to take a magnesium supplement on a regular schedule and increase the dosage when you are stressed, or when you are about to take part in a stressful activity.

MAGNESIUM AND HEART ATTACKS

Ideally, of course, we would all modify our diets and lifestyles (and arrange to be born into families with no history of heart disease) so as to eliminate the danger of heart attack forever. In the real world, however, about 1.5 million people in the United States suffer heart attacks every year, and roughly a third of them die. Obviously, this is a serious health problem that is not going to disappear overnight. But that does not mean there is no hope. Research shows that many heart attacks can be prevented. And when heart attacks do occur, many of the deaths that result might possibly be prevented also, if we added a simple natural element—magnesium—to the standard treatment.

Research demonstrates that heart attacks cause a depletion of the magnesium in heart muscles. Analysis of tissues from the hearts of people who have died from heart attacks shows magnesium levels of 42 to 50 percent below normal in the dead tissue, and 19 to 27 percent below normal in the tissue that did not die, but was starved for blood during the attack.[6] Those with fatal heart seizures caused by ischemia lost 12 to 22 percent of the normal magnesium concentration in the cells of the heart muscles.[7] People who died suddenly of heart failure had magnesium deficits of 12 to 16 percent.[8]

Dr. A.S. Abraham and his coworkers studied the release of magnesium during heart attacks in rats.[9] Their data show that when a heart attack occurs, there is an initial loss of magnesium from the heart muscles. This levels off to a stable, but deficient, status in about half an hour. As the heart muscle loses magnesium, the magnesium concentration in the blood serum shows a marked increase. It reaches a peak in about one hour, then gradually tapers off to a 10-percent deficit within twenty-four hours of the beginning of the attack.

Dr. Abraham and J.R. Marier tell us that the magnesium expelled from the heart muscle migrates into the blood, from which it is eliminated through the kidneys. In a more recent report, Dr. Henrik S. Rasmussen of the Hvidivre Hospital at the University of Copenhagen, Denmark, suggests that rather than being excreted, the released magnesium reacts with free fatty acids in the blood to form an inactive chemical complex.[10] Whatever the case, the magnesium is unavailable to perform its normal functions within the cells of the heart muscle. Over time, if the victim survives the attack, the magnesium deficit progressively disappears as active magnesium is restored through the ingestion of food and water.

This strongly suggests that patients who have recently suffered a heart attack should receive immediate treatment to maintain a normal level of magnesium in the heart muscle. And, in fact, there are two methods of doing this that have shown positive results.

In the first, reported by Dr. Y. Rayssiguier, the infusion of a salt (sodium nicotinate, which is a compound of sodium and niacin) prevented an increase in the concentration of free fatty acids in ewes that had also been infused with adrenaline.[11] As a result, the normal, active form of magnesium in the cells remained essentially unchanged. Similarly, Dr. M.J. Rowe and colleagues administered a compound similar to nicotinic acid to patients during heart attacks to reduce the incidence of ventricular arrhythmias.[12] It was effective, because the compound inhibited the formation of magnesium-stealing free fatty acids that can be caused by the catalytic action of adrenaline, and the magnesium concentration in the muscle cells of the patients' hearts remained relatively constant.

The other method of maintaining magnesium levels in heart attack patients simply involves the intravenous infusion of a magnesium compound to restore the magnesium balance as soon as possible. Dr. B.C. Morton and his coworkers at the University of Ottawa in Ontario, Canada, have shown that such an infusion of magnesium soon after the onset of a heart attack can reduce the size of the area of heart muscle that suffers permanent damage.[13] They worked with patients who were having their first heart attacks. The doctors also noted a reduction in the occurrence of ventricular arrhythmias.

Based on these findings, and extrapolating from the research of Dr. I.H. Chaudry and his coworkers (see page 63), it seems reasonable to suggest that the intravenous infusion of Mg:ATP could be an effective way to help inhibit the death of heart tissue during a heart attack. Recall that infusions of the chemical compound Mg:ATP increased the heart function of healthy volunteers by an average of 131 percent without increasing blood pressure. That, my friends, is heart power!

Dr. Rasmussen divided 130 patients with acute myocardial infarction into two groups.[14] One group, consisting of 74 patients, received a placebo; to the other group, comprising 56 patients, he administered magnesium. During their first forty-eight hours in the hospital, the magnesium-treated group received 1.5 grams (1,500 milligrams) of magnesium ion from 5.9 grams of magnesium chloride in an intravenous glucose solution. The dosage thus amounted to 750 milligrams of elemental magnesium per day. The intravenous glucose fluid administered to the placebo group contained no magnesium. During the course of treatment, the patients did not know which group they were in.

In the first four weeks of hospitalization, 19 percent of the placebo group (fourteen patients) died—half of them from shock treatments administered to stop seizures of arrhythmia. In the same time period, 7 percent in the treatment group (four patients) died, none of them as a result of shock treatment. In all, 47 percent of the placebo group (thirty-five patients) needed shock treatment for arrhythmias, as compared with 21 percent in the magnesium group (twelve patients).

Dr. Rasmussen informs us in the same report that in a larger group, comprising 273 patients, 136 received magnesium infusion and 137 were given a placebo. Forty-one percent of the magnesium-treated group (56 patients) developed acute myocardial infarctions as a result of their heart attacks, as compared with 54 percent in the placebo group (74 patients). Thus, he demonstrated that magnesium protects against heart muscle damage as a result of heart attacks. He states in his concluding remarks that magnesium treatment should be adopted as part of routine clinical practice for patients with acute myocardial infarction. He is currently conducting a study on the effects of oral supplementation of magnesium in patients who have survived serious heart attacks. The results should be interesting.

Blood-vessel blockage, arrhythmia, and congestive heart disease all deny the heart the nutrition it needs to operate efficiently. Severe curtailment of its blood supply induces the heart to decrease its pumping action to a minimum level. When this happens, the patient experiences the symptoms of a heart attack. If the heart's blood supply is reduced to a low enough level, the heart ceases to work at all—the patient goes into a condition of cardiac arrest. Unless immediate revival attempts succeed, the patient dies.

Not all heart attacks are the result of arterial blockage. A different, though less common, heart attack scenario takes shape when a weak section called an aneurysm develops in an artery. This expands in size at the point of weakness, which causes the arterial walls to become thinner and thinner. At some point, the artery's wall ruptures, and massive internal bleeding occurs. If the ruptured artery is large and the bleeding that results is massive, the circulation of blood will terminate, and the heart (along with all the other parts of the body) will die from lack of nutrition. Unfortunately, aneurysms are most often the result of a congenital defect, so there is usually little or nothing that can be done to prevent them, although if they are detected before they rupture, in some cases they can be repaired.

STROKE

As physicians would phrase it, a heart attack is one "primary end point" to cardiovascular disease. Stroke is another. Some strokes are mild; some are severe. Many are fatal. Usually, the victim of a stroke loses enough brain function to become at least partially paralyzed. Some people lose their ability to talk; others become unable to walk.

Several years ago, I attended my fiftieth high school reunion. I sat at dinner beside a classmate who had been restored to health by heart bypass surgery; but I was shocked and saddened to see two men in wheelchairs. They were the victims of strokes. They had been sentenced to spend the rest of their lives as invalids. One was unable to walk; the other could neither walk nor talk. The former experienced his stroke during heart bypass surgery. During such operations, blood clots sometimes become dislodged and travel to the brain, where they cause strokes. This is a very real danger during bypass surgery.

Recently I met a lady who walked, haltingly, with a four-footed cane. Her right side functioned normally, but she had suffered partial paralysis of her left leg and total paralysis of her left arm. For such a patient, physical therapy can in some cases be helpful, but if the brain is severely damaged, there is little anyone can do.

These examples from real life serve to impress upon us the severity of strokes and the effect they have on those who survive them.

WHAT CAUSES STROKES?

When the blood flow to a portion of the brain becomes so restricted that a group of brain cells ceases to function, a person is said to have had a stroke. The extent and location of the stroke determines the damage it does to total brain function.

Strokes are generally classified as either ischemic or hemorrhagic, depending on the events that induce them. Many strokes occur as a result of a blockage that stops the flow of blood to some part of the brain. Such a seizure is called a *cerebral thrombosis*, or an ischemic stroke. Hemorrhagic strokes occur as a result of the rupture of a blood vessel. This causes a hemorrhage, which, in turn, deprives a portion of the brain of its supply of blood.

A third cause of strokes, not as commonly discussed, is the occurrence of a spasm in one or more of the brain's arteries. In this scenario, one or more blood vessels in the brain—which may or may not have existing blockage—undergoes a spasm that restricts or even blocks the flow of blood, causing a portion of the brain to become starved for oxygen. If such an attack is relatively mild—with symptoms ranging from momentary confusion to a brief loss of consciousness—and subsequent recovery is complete, it would likely be classified as a *transient ischemic attack*. A transient ischemic attack is a stroke in which the blood supply to a portion of the brain is temporarily restricted to dangerous levels, but is restored before permanent damage to the brain cells can take place.

Among other things, stress can induce the arteries in the brain to constrict in spasms. Spasms in the blood vessels of the brain are like those in other blood vessels. They are commonly caused by a deficiency of magnesium, which induces magnesium and potassium to transfer out of the smooth muscle of the blood vessels. Calcium and sodium then move in to take their place. Excess calcium and sodium in the smooth muscles cause the blood vessels to develop tone and constrict, narrowing their inner diameters. As a result, blood pressure increases and blood flow decreases. This can cause one's mental abilities to freeze up. It is probably the physical cause of stage fright.

If a spasm is mild, it may be temporary, reversible, and cause no permanent damage. If it is more severe, it can lead to any degree of injury, including death. Drs. Burton and Bella Altura have been doing research in this area for a number of years, and have written many scientific papers on the positive effect of magnesium on the health of blood vessels in all parts of the body. They have done much of their research on animal subjects, but, because animals' blood vessels work much like our own, their data translate directly to the behavior of human blood vessels under the same conditions.

Dr. R.S.A. Tindall reveals a frightening statistic about the danger of mild strokes. He says that 30 to 40 percent of those patients who have more than one transient ischemic attack will, within the next five years, have a cerebral infarction (a severe stroke resulting in dead brain tissue), with only a 50-percent chance of survival.[15] Thus, even mild ischemic events, including cerebrovascular spasms, can foretell future disaster.

MAGNESIUM AND THE BRAIN

As it is in the heart, magnesium is a principal guardian of healthy cells in the brain. In fact, it is present in a higher concentration in the cerebrospinal fluid (the fluid that bathes the spinal cord and brain) than in the blood plasma.

Working with animal subjects, Drs. Bella and Burton Altura measured the tension in arteries in the brain in the presence of different degrees of magnesium deficiency. They found that decreasing the concentration of magnesium in the cerebrospinal fluid contacting the arteries caused the degree of tension in the arteries to increase. The lower the magnesium level, the greater the tension was.[16] This means that the lower the concentration of magnesium in the brain, the greater the risk of arterial spasms.

From what we know about the action of magnesium in the body, we can conclude that it protects against ischemic strokes by preventing spasms in the brain's circulatory network. Magnesium also helps to prevent the abnormal formation of blood clots because, as we have seen, it inhibits the deposition of blood platelets and increases the fluidity of blood. Because it acts to reduce blood pressure and maintain it at a normal level, magnesium also protects against hemorrhagic strokes.

I cannot guarantee, of course, that a person who takes magnesium supplements will never have a stroke. But the data strongly imply that the risk of having either a hemorrhagic or ischemic stroke will be minimized if you maintain a normal magnesium level in your tissues. My personal experience with taking magnesium supplements has led me to banish all fear of ever having a heart attack or stroke as long as I am able to maintain a healthful level of minerals and vitamins.

In order to maintain this balance, we need to eat significant amounts of grains and vegetables. This is the essence of the diets developed by Michio Kushi, Nathan Pritikin, Dr. Julian Whitaker, and Dr. Dean Ornish.[17] Everyone should monitor his or her nutritional health by having a periodic blood analysis performed at a reputable laboratory. Consult with your physician for recommendations regarding medical laboratories that perform this service, because an erroneous analysis is worse than none at all.

Both stroke and heart attack can result from severe ischemia. Regardless of the immediate cause, when either the brain or the heart receives too little blood to operate effectively, the organ diminishes its functions to a level commensurate with the amount of blood it receives. If either organ receives insufficient blood to operate, the person dies. In between normal health and death lies a spectrum of stages of debility.

If we wish to avoid the consequences of either a heart attack or stroke, we must adopt a lifestyle that avoids abusing our bodies with unnecessary degrading habits—smoking, excessive alcohol consumption, taking drugs, getting too little sleep, exposing ourselves unnecessarily to stress, and more. You can fill in the blanks for your own life. We must also nourish our bodies in a healthful manner. You can do this by adjusting your diet to eat healthfully, and by adopting an appropriate program of nutritional supplements to make sure that you maintain a healthy balance of essential nutrients—among which magnesium dominates—at all times.

6

An Aspirin a Day?

By this time you are probably thinking, "I hear what you're saying about magnesium, but what about aspirin? Almost everyone I know who has heart trouble, or has survived a heart attack or stroke, is taking aspirin. Even the doctors are taking it. Shouldn't I?"

Aspirin or magnesium—which is best? Do we need both? If not, who should take aspirin? Who should take magnesium? What evidence is there that taking *either* aspirin or magnesium on a daily basis really helps? What are their respective benefits? What are the risks? How much of either should we take? In short, how do aspirin and magnesium compare as agents for preventing and treating cardiovascular disease?

In the previous chapters, we have explored magnesium's ability to prevent and minimize the effects of cardiovascular disease. But we have not said much about aspirin, which is now very commonly prescribed as a preventive measure for people at risk of heart disease. So let us now take a look at the evidence physicians use to support this practice.

First of all, we need to consider the logic behind using aspirin to prevent heart attacks. One of aspirin's notable properties is that it inhibits the formation of blood clots in the arteries. It does this by stopping the production of chemicals that cause blood platelets to be sticky.[1] And if blood platelets can't stick together and to the walls of the arteries, they can't form clots.

The danger of having clots form in the arteries, as we have seen, is that they slow down and sometimes stop the flow of blood. If blood clots form in the arteries that feed the heart, they may cause a heart attack. In the brain, they can cause a stroke. Clots tend to form in arteries that are already partially blocked, and aspirin is frequently prescribed to inhibit their formation.

Aspirin effectively inhibits the action of cyclooxygenase, which is the enzyme responsible for making prostaglandins. Some of these prostaglandins cause blood platelets to adhere to one another and to the walls of blood vessels. But other prostaglandins do exactly the opposite—they inhibit platelet cohesion and the adhesion of platelets to arterial surfaces. Thus, aspirin destroys the body's natural ability to inhibit the formation of clots and substitutes its own clot-controlling action.

In a healthy body, blood platelets adhere to arterial walls when they are needed to form a repair patch in the artery's lining; otherwise, they remain harmlessly in the bloodstream. Prostaglandins are hormonelike substances that regulate the balance between the opposing tendencies (to clot or not to clot). Aspirin, which prevents the body from producing these prostaglandins, inhibits both of these rival processes.[2]

Second, before we turn to the research on the effectiveness of aspirin, we need to know how to look critically at the reported findings so as to avoid being deceived. When studies are done to evaluate aspirin's direct effect in reducing the incidence of heart attacks and strokes, no advantage can be found among small numbers of patients. Such studies need to include information about large groups of people collected over extended periods of time. Then the results are analyzed by sophisticated computer programs to search for statistically significant results. This is necessary because in all of the aspirin studies, benefits appear in only a small percentage of the population being tested. In other words, with the aspirin studies, it is impossible to determine that any particular individual is likely to benefit from taking the medication. Any benefits referred to are strictly statistical.

Then, within the small group of people who are statistically determined to have been helped, investigators may calculate benefits amounting to upwards of 50 percent. This sounds impressive, but we must understand that what this actually refers to is not 50 percent of a study's participants, but 50 percent of only 1 to 2 percent of the *total* sample of people. The small numbers in these cases are never quoted, but they are the real percentages of the population of each study that benefit from taking aspirin. Obviously, if only 1 or 2 percent of a group of people receive a benefit, that means that 98 or 99 percent get no benefit at all. We shall return to this issue when we look at specific studies.

Finally, the aspirin studies routinely use questionnaires to collect data. They therefore assume that all participants report their variables correctly. If questionnaires are sent out on an annual basis and some of the participants keep poor records, the validity of the reported data will be no better than the participants' ability to remember the events of the past year.

With those cautions aside, let us turn to the study results.

WHAT THE ASPIRIN RESEARCH TELLS US

In a study done in the United States and published in 1975, one million men and women reported in questionnaires on the effect aspirin had on the incidence of coronary heart disease among them. The results showed no effect.[3]

In another study, the Coronary Drug Project Aspirin Study, a total of 1,529 men who had had a heart attack were observed for three years, with ten to twenty-eight months of follow-up afterward. Half took 324 milligrams of aspirin daily; the others took a placebo. During the period of observation, 5.8 percent of those taking aspirin and 8.3 percent of those in the placebo group died.[4] The results indicated that aspirin may have had a positive effect, but because of errors inherent in the data, a statistical analysis indicated that no definite conclusion could be reached.

Dr. Charles Hennekens and his coworkers did a study in which they read the death certificates and interviewed the widows of 568 married white men in two Florida counties over a sixteen-month period. They found no evidence that aspirin gave any protection against heart disease.[5]

The National Heart, Lung, and Blood Institute conducted a study of persons thirty to sixty-nine years of age, all of whom had survived heart attacks. They were enlisted from thirty clinical centers in the United States. Of the 4,524 participants, 2,267 took one gram of aspirin per day and 2,257 took a placebo. During that thirteen-month study, 9.6 percent of the aspirin group and 8.8 percent of the placebo group died. After the study was concluded, the patients' progress was followed for an additional three years. Over the entire period, 10.8 percent in the aspirin group died, as did 9.7 percent in the placebo group. The total incidence of heart attacks among the two groups was 14.1 percent and 14.8 percent, respectively. The researchers also tracked the side effects experienced by study participants. Severe stomach irritation occurred in 23.7 percent of those taking aspirin and 14.9 percent of the placebo group. These results led to the conclusion that aspirin should not be recommended for patients who have already had a heart attack.[6]

The Swedish Angina Pectoris Aspirin Trial, published in November 1992, is the most recent study on the effect of aspirin on the health of patients with the classic form of chest pain, stable angina pectoris. The study involved 2,035 patients in 94 treatment centers. They were under the care of 108 different physicians. The median age of the patients was 67 years; 52 percent of them were male. (Just think about the possibility for error—there were 2,035 participants, 94 different treatment centers, and 108 different physicians!)

Aspirin was found to have a beneficial effect on "primary events" (defined as fatal or nonfatal heart attack [myocardial infarction] and sudden death) in just 2 *percent* of the total sample. The effect aspirin had on "secondary events," which included strokes, blood vessel failure, and death from all causes, showed a benefit to 5.2 percent of the patients.[7] And this is an unusually *high* percentage of people to receive a benefit from taking aspirin.

THE STUDY THAT MADE HEADLINES

A number of other studies have been done, but the one that seems to be viewed as the standard for measuring the success of aspirin in reducing the incidence of heart attacks and strokes is the Aspirin

Component of the Ongoing Physicians' Health Study.[8] At least, this is the one that caused the recent surge in the use of aspirin. This investigation was organized and conducted in the United States by the Steering Committee of the Physicians' Health Study Research Group, led by Dr. Charles Hennekens, the principal investigator. The study was designed to observe, among other things, the effect of aspirin on the risk of a person having a first cardiovascular event. Because of the enormous impact this report has had on the medical community and the patients they treat, you need to know some of the details about it.

How the Study Was Done

First, extensive preliminary screening eliminated individuals who had in their medical history any one of the following: heart attack, stroke, transient ischemic attack, cancer (except nonmelanoma skin cancer), current liver or kidney disease, stomach ulcer, or gout. Also eliminated were people who were sensitive to aspirin and people who were currently taking aspirin, other platelet-active drugs, nonsteroidal anti-inflammatory agents, or vitamin A supplements. This reduced a sample of 59,285 willing participants to a list of 33,223 who were eligible. An initial trial run of eighteen weeks decreased that number further, to a final selection of 22,071 male physicians ranging in age from forty to eighty-four.

The test subjects were divided into four groups of equal size, and placed on the following regimens:

- Group A took 325 milligrams of Bufferin every other day, and 50 milligrams of beta-carotene on the alternate days.
- Group B took 325 milligrams of Bufferin every other day, and a placebo in place of beta-carotene on the alternate days.
- Group C took a placebo in place of Bufferin every other day, and 50 milligrams of beta-carotene on the alternate days.
- Group D took placebos in place of both the Bufferin and beta-carotene.

As a result, a total of 11,037 participants took Bufferin and 11,034 received a placebo. It should be noted that although the project was called an aspirin study, Bufferin is not pure aspirin. Bufferin tablets contain aspirin together with the compounds calcium carbonate and magnesium oxide, which are added to moderate the acidity of the stomach and reduce the stomach irritation aspirin can cause. Aspirin is a unique chemical compound (acetylsalicylic acid). Bufferin is a mixture.

Two reports of the study's results were published. In the first, the preliminary report, the average follow-up time was 57 months; in the second, the final report, it was 60.2 months. The preliminary report tells us that, of all the 22,071 participants—those who took Bufferin and those who took a placebo—88 deaths due to cardiovascular events occurred in 4.8 years.[9] These deaths were evenly divided, with 44 occurring in each of the two groups. That amounts to 0.4 percent of the total sample of participating physicians.

What the Results Mean

On the simplest level, to determine the benefit of Bufferin, one would look at the difference between the number of deaths that occurred in the Bufferin group and the number that occurred in the placebo group. In this case, therefore, since there were 44 deaths in the Bufferin group and 44 among the placebo group, the net benefit was zero.

Further, statistics show that among 22,071 American men from forty to eighty-two years of age, the average number of deaths in a 4.8-year period is 733, or 3.3 percent.[10] This is significantly higher than the 0.4 percent observed in the study. Even if we were to attribute the lower death rate to the Bufferin, we would still expect somewhere between 300 and 400—not 44—deaths in the placebo group. The Bufferin group did suffer fewer deaths from heart attack than the placebo group, but that reduction was exactly offset by a higher incidence of deaths from strokes, ischemic heart disease, sudden cardiac death, and other cardiovascular and cerebrovascular events. The total number of deaths in the two groups was the same.

Because the number of deaths in the placebo group was as low as that in the Bufferin group, we can rule out any significant role for aspirin in the prevention of fatal heart attacks or other fatal cardiovascular events among this sample of physicians. What these data compel us to conclude is that for some reason (perhaps the extensive preliminary screening or a healthier than average lifestyle) the physicians who participated in the study were not a representative sample of American men. Even in the placebo group, the number of deaths was only 12 percent of what would be expected among that many men in the general population. On that basis alone, the study's conclusions should be disregarded as being unreliable. Extrapolating from data obtained by observing 22,071 healthier than average physicians to predict the effect of aspirin on the much less healthy general population is certainly questionable.

The major benefit cited in the reports on the study was a reduced risk of heart attack. This was given as a 47-percent reduction in the preliminary report and a 44-percent reduction in the final report.[11] The difference between those two numbers means that there is a 6.4-percent deviation in a single statistic, obtained from the same study, using data that applies to the same starting date, with end points only 3.2 months apart. Somehow, a figure obtained after a nearly five-year study dropped by 6.4 percent in a few months. This discrepancy is puzzling, and the authors leave it unexplained.

Going beyond that statistical discrepancy, it would seem that these numbers mean that between 44 and 47 percent of the physicians who took Bufferin received some benefit from taking it. But that is not the way the system works. In the final report, there were 139 heart attacks in the Bufferin group and 239 in the placebo group. If we attribute the difference between these two numbers to the effect of the Bufferin, we would conclude that 100 heart attacks were prevented by Bufferin. Thus, the fraction of the total group that benefited is 0.00906 (the 100 people who benefited divided by 11,037, the total number of Bufferin-takers), or 0.906 percent. In other words, slightly less than 1 percent benefited. This means that over 99 percent of the people in the Bufferin group received no benefit at all.

This is the number they don't talk about. *Only nine-tenths of 1 percent of the participants in the Bufferin group actually experienced a lowered risk of having a heart attack.* So how did the researchers produce the calculation of a 44-percent (or 47-percent) reduction of risk? They did it by taking at the number of people who presumably didn't have heart attacks because of taking Bufferin (100) and looking at it as a percentage of the number of people who did have heart attacks and didn't take Bufferin (239). So the 0.9 percent of the Bufferin-taking participants who benefited from aspirin are the same people who are said to represent a 44-percent *risk reduction* among those who had heart attacks. You have just witnessed some of the magic of statistical analysis.

Another finding that received little notice regarded the incidence of various types of strokes. The final report on the study showed a higher incidence of stroke among those who took Bufferin than among those who didn't.[12] Specifically, there were 119 strokes in the Bufferin groups and 98 in the placebo groups. In the preliminary report these numbers were 80 and 70, respectively.[13] Consider again the two major causes of stroke: obstruction in a blood vessel in the brain, and a rupture in a blood vessel that results in hemorrhage. Either of these two conditions deprives brain cells of the blood supply they need to function and live.[14]

Supposedly, strokes caused by blockage (in which, as we have seen, arterial spasms may play a role) can be prevented by the use of aspirin or some other thrombolytic (clot-preventing) agent. (On the other hand, internal bleeding—and presumably, therefore, some hemorrhagic strokes—can be *caused* by such agents.) Yet the Final Report of the Aspirin Component of the Ongoing Physicians' Health Study tells us that the Bufferin group had more strokes caused by blockage (91) than did the placebo group (82).[15] In the preliminary report, these numbers were 64 and 61, respectively, with one stroke of unknown origin in the Bufferin group. That's right! The Bufferin group had *more* ischemic strokes than the placebo group. Yet physicians tell us that aspirin is supposed to prevent that kind of stroke.

Meanwhile, as one might expect, Bufferin seemed to be implicated in the development of ruptured blood vessels leading to strokes. There were 23 hemorrhagic strokes among people taking Bufferin, but only 12 in those taking a placebo. In this case the numbers are the same in both reports. Ultimately, what these data tell us is that Bufferin has no beneficial effect in reducing the effect of either kind of stroke.

Statistically, these results are insignificant. But the same pattern is found in other studies.[16] And if we follow the lead of the report's statisticians and consider only the small group who had strokes, we find a 21.4-percent *greater* incidence of strokes among those who took Bufferin than among those who did not.

One of the study's conclusions was that, before prescribing aspirin, physicians should check their patients to determine whether they are more likely to be susceptible to an ischemic or a hemorrhagic stroke. This seems like a wise suggestion, but in reality it would probably be of dubious value, since aspirin seems to be associated with an increase in both kinds of strokes.

Several other conclusions were presented in the report, but the final one is the most significant. This is that, in terms of primary prevention, *Bufferin caused no decrease in risk of death from all cardiovascular causes.* At this point, the reported 44-percent reduction in risk of a heart attack melts away. We have to ask why physicians insist on prescribing aspirin as a therapeutic agent. If there is no net benefit, what is the value of it? Who profits—the patients, or the drug companies and physicians?

One other clear finding, based on the fact that the incidence of heart attack and stroke was much lower than expected among all study participants, is that the 22,071 physicians studied were decidedly much healthier than the average American male. Also, the authors tell us that the study would have had to continue for another ten years (for a total of about fifteen years) for them to obtain results that show clearly what the effect of aspirin on death from all cardiovascular causes really is. We must finally conclude that the beneficial effects of aspirin on this sample of men were, at best, quite small.

What About the Beta-Carotene?

So far, we have discussed the study's findings only as they relate to Bufferin. More recently, Dr. J. Michael Gaziano of the Harvard Medical School reported an illuminating discovery. Recall that half of the 22,071 physicians who participated in the Ongoing Physicians' Health Study took 50 milligrams of beta-carotene every other day, while the other half took a placebo. In these two groups were 333 physicians with symptoms of coronary artery disease. When the data were statistically adjusted to allow for any effects of taking Bufferin, those who took beta-carotene were found to have had only half as many cardiovascular events as those who did not. This portion of the study extended over a period of six years.[17]

Beta-carotene is a natural nutrient, and the 50-percent risk reduction it produced is greater than any reported for Bufferin. And it had this effect whether or not the participants also took Bufferin. Although the sample size is small, it seems reasonable to conclude that the data must reflect the beneficial effect caused by beta-carotene.

Research indicates that beta-carotene inhibits the oxidation of cholesterol and low-density lipoprotein (LDL), decreasing their tendency to form plaque.[18] The body converts beta-carotene into vitamin A, and both beta-carotene and vitamin A neutralize toxic free radicals, which contribute to the development of various diseases (see Chemical Free Radicals, page 93).

So this study—which was intended to prove that aspirin prevents death from cardiovascular disease—ended up showing that the most beneficial effect was due to a natural nutrient, beta-carotene. This must have been surprising to the investigators, who had included beta-carotene in the study to see if it had any value in preventing cancer, not heart disease.

Proven, effective nutritional therapies exist, but are in very limited use by the medical profession. Yet nutritional therapies convey all the benefits for which aspirin is touted, and more, and do so much more effectively and with no side effects. Not only do they make many sick people better, they make them *well.*

Dr. R.A. Riemersma of the Department of Cardiology and Medicine at the University of Edinburgh, Scotland, and his colleagues found that the well-known antioxidants vitamin C, vitamin E, and beta-carotene can reduce the incidence of chest pain caused by angina.[19] In this application, the most effective of these agents is vitamin E. Although vitamin C is a stronger antioxidant overall than vitamin E, it is apparently less effective in controlling angina pectoris. Beta-carotene was the least effective of the three for this purpose.

In 1966, Herbert Bailey reviewed what was then known about vitamin E and cited the pioneering work done at the Shute Institute

Chemical Free Radicals

Many degenerative diseases are associated with aging and are not caused by germs or viruses. These include atherosclerosis, high blood pressure, heart attack, stroke, macular degeneration, cataracts, adult-onset diabetes, and cancer. In most cases, these diseases are in some way caused or exacerbated by reactions involving chemical free radicals in our bodies.

Free radicals can be atoms, ions, molecules, or fragments of molecules. The one thing they have in common is that they contain at least one electron that is "free"—that is, not paired with another electron.

Atoms generally consist of dense, positively charged nuclei surrounded by systems of electrons, which carry a negative charge. The electrons usually occur in pairs that form a stable arrangement. When an atom has an odd number of electrons, however, one of them obviously cannot be part of a pair. A chlorine atom, which has seventeen electrons, is an example. It has eight pairs of electrons and one unpaired electron. This is the simplest kind of chemical free radical.

Groups of atoms react together to form molecules. They do this by sharing electrons between them. If two atoms each have one unpaired electron, these can form a pair that is attracted to both atomic nuclei, thereby creating a molecule that is chemically related to the two original atoms, but also different.

Because atoms and molecules always strive for stability, an unpaired electron is always going to be, in effect, looking for another electron to become paired with, and will readily enter into a chemical reaction when the opportunity presents itself. This is what makes free radicals, in chemical terms, unstable and reactive.

It is important to distinguish destructive free-radical reactions from nondestructive ones. Not all free-radical reactions are harmful. In fact, many of them are necessary to sustain life. For example, the oxygen we breathe into our lungs some 25,000 times each day is a chemical free radical. This is what makes it possible for oxygen molecules to bind with the iron atoms in the hemoglobin of red blood cells. The hemoglobin transfers the oxygen to other oxygen-carriers by way of a series of complex biochemical reactions and eventually delivers the oxygen to the sites at which the body produces energy from the oxidation of glucose. The culmination of this sequence of events is the formation of carbon dioxide and water as the end products that are eliminated principally through the lungs and kidneys.

The chemical free radicals that can do so much harm are the byproducts of a number of different processes and have many different chemical identities. Many are generated by the body in the course of normal metabolism. As long as the body's chemical balance is normal, it utilizes free radicals as quickly as they are formed. But when there are too many free radicals for the body to cope with, they can become toxic. And because they are so reactive chemically, they can also get involved in such mischief as the formation of plaque in the arteries, the acceleration of the aging process, and the generation of cancerous tissues, among other things.

There are several reasons why many of us are beset with the problem of too many free radicals. First of all, the high level of fats and oils in our diets causes the formation of many more free radicals than the body needs to conduct its normal chemical business, especially when we use these substances for frying. Deep-frying is particularly bad. (Think of this the next time you find yourself about to order French fries or potato chips.)

In addition, our modern, industrialized environment literally bombards us with free radicals, among them ozone (created by electric motors and smog) and nitrogen oxides (a product of the combustion of hydrocarbon fuels in internal combustion engines and furnaces). Even worse is the incomplete combustion of diesel and jet fuels, which forms complex compounds known as polycyclic aromatic hydrocarbons. These are easily transformed into ion radicals—charged free radicals—many of which are potent cancer-forming species. Tobacco smoke is another source of dangerous free radicals.

X-rays and radiation therapy break chemical bonds and generate free radicals. These are highly reactive chemical species that immediately react with anything nearby. None of the products of these reactions is likely to be a compound that is useful to the body. Even the local water supply can be a source of toxic free radicals. Because many municipalities add chlorine to the water to kill bacteria, Americans consume large quantities of chlorinated water. The chlorine and residual organic compounds dissolved in water can react together to form cancer-inducing compounds.

The reactions of free radicals in the body are very complex. They form and break apart in breathtaking ways. As a result, although they have been implicated in many degenerative disease processes, the exact mechanisms by which they create or worsen diseases are not always precisely understood. It is known, however, that a number of different agents act as free-radical trapping agents (which destroy the free radicals) and antioxidants (which keep them from reacting in dangerous ways in the body). There are others, but among the most potent of these are the old standbys beta-carotene and vitamins A, C, and E. This is one reason why proper nutrition is essential for the prevention of disease and the maintenance of good health.

in London, Canada, more than ten years earlier.[20] The beneficial effects of vitamins C and E, along with a number of other nutrients, have been discussed and supported by Dr. Roger J. Williams, one of the greatest of modern biochemists.[21]

Vitamin E unloads the heart. This means that it decreases the amount of oxygen the heart needs to do its work. It makes the heart work more efficiently. My own experiences with taking vitamin E totally support this statement. Magnesium also unloads the heart. The combined actions of magnesium and vitamin E in unloading the heart are nothing short of miraculous. The action of vitamin C combined with vitamin E and magnesium is even better. In contrast, I am not aware of any report claiming that aspirin makes the heart work more efficiently.

Before leaving this aspect of the Ongoing Physicians' Health Study, we must also ask how many of the participants were taking vitamins C or E during the course of the investigation. Those taking vitamin A were excluded from the study, even though beta-carotene, a component of the study, is converted to vitamin A as the body needs it. The study required half of the participants to take beta-carotene. How many were taking mineral supplements, perhaps magnesium or potassium? What were the dietary habits of the participants? These matters were not mentioned in the reports, but they are very important in determining the validity of the results.

THE ENGLISH ASPIRIN STUDY

In England, a study involving 5,139 physicians extended from November 1978 to November 1984. This study was not totally independent of the American study, because its investigators included three who were involved in directing the Ongoing Physicians' Health Study in the United States. They were Drs. Charles Hennekens (from the United States), Richard Peto (Britain), and Richard Doll (Britain). The study was an open one involving no placebos. Two-thirds of the physicians took either 500 milligrams of regular aspirin, which is absorbed in the stomach, or 300 milligrams of coated aspirin that is absorbed in the intestines. During the course of the study, some of the participants stopped taking the aspirin, so that at the halfway point about 70 percent of those originally assigned to take aspirin were still doing so. Note that in this study the participants took over three times as much aspirin in a given period as the physicians in the American study did.[22]

The British study showed that no statistically significant benefit was realized. There was a 10-percent lower incidence of death in the aspirin group as opposed to those who took no aspirin. This fact allows for a possible (but unconfirmed) beneficial effect of 10 to 20 percent in the aspirin group. However, no significant difference was found between the incidence of heart attacks in those who took aspirin and those who did not.

The proportional number of minor strokes was only half as great in the aspirin group as among those who took no aspirin. This sounds like good news, until you combine it with the the fact that the *total* number of strokes in the aspirin group was higher than the number in the control group. In other words, those taking aspirin had a lower incidence of minor strokes but a higher incidence of major, disabling strokes. This is a concern because it may indicate that aspirin tends to cause strokes to be more severe than they would otherwise be. It follows a pattern similar to that found in the American study. As in the American study, *there was no reduction in overall deaths from all cardiovascular causes as a result of taking aspirin.*[23]

BENEFITS OF ASPIRIN FOR SECOND EVENTS

So far we have been discussing the effects of aspirin in preventing first events—the initial occurrence of heart attack or stroke. An analysis of the effects of aspirin on patients who have already had one or more strokes or heart attacks yields different results. A team of British researchers reviewed and statistically correlated the findings of twenty-five separate studies conducted for the purpose of evaluating the benefits of aspirin and other anticlotting medications.[24] They found that no other medication used to inhibit the formation of blood clots was any more effective than aspirin. Aspirin also had a positive influence on the survival rate of patients whose disease was caused by blockage of blood vessels.

The report's final conclusion states that on a statistical basis, aspirin is beneficial in treating cardiovascular diseases in patients who have a history of one of those diseases. Thus, if you have suffered a stroke or heart attack, the report suggests that aspirin may reduce your chance of having a fatal attack by about one-sixth, and your chance of experiencing a nonfatal event by about one-third.

However, a dark cloud still shadows the use of aspirin, even for second events. In a recent report in the journal *Thrombosis Research*, Dr. K.-H. Grotemeyer of the Department of Neurology at the University of Münster in Germany discussed his observations on aspirin's effectiveness at preventing platelets from forming blood clots. When hospitalized patients suffering from strokes were given 500 milligrams of aspirin, 10 percent of them showed enhanced—not reduced—platelet activity within two hours. After twelve hours, 36 percent of the patients had enhanced platelet activity.[25]

In the November 1993 issue of his newletter *Alternatives for the Health Conscious Individual*, Dr. David G. Williams also refers to Dr. Grotemeyer's research. Dr. Grotemeyer has found that approximately 33 percent of his carotid stroke patients who took aspirin on a continuing basis for two years experienced a rebound reaction that caused an *increased* risk of their developing blood clots in their arteries. During that time period, 4.4 percent of those who responded favorably to aspirin died of cardiovascular events, but *40 percent* of those who suffered rebound reactions died. Thus, for a substantial portion of his sample, taking aspirin resulted in an 89-percent higher likelihood of cardiovascular death.[26]

In addition to his own research, done on real patients in a

clinical setting, Dr. Grotemeyer cites a number of other research reports that likewise show aspirin to be ineffective in preventing strokes. The discrepancies between these data and those reported in the major aspirin studies need to be investigated, and this should be done by disinterested researchers whose work is not funded by the big drug companies. We deserve to know the truth about the long-term use of aspirin.

GOOD NUTRITION IS BETTER THAN ASPIRIN

When all things are considered, however, the use of a nutritional therapy to prevent cardiovascular disease is still preferable. The success rate among people who faithfully engage in nutritional therapy is so high that there is no need to perform studies with large numbers of people to see its benefits. Nor is a statistical analysis of a large amount of data required (although some recent studies have included statistical analyses of their data, presumably to give them credibility among medical professionals).

Physicians at the Tufts University School of Medicine in Boston, Massachusetts, and the Johns Hopkins University School of Medicine in Baltimore, Maryland, used statistical methods to analyze data on a study of 100 patients who had undergone heart surgery.[27] After the cardiopulmonary bypass operations were completed, 50 of the patients received an intravenous infusion of two grams of magnesium chloride (or a total of 814 milligrams of elemental magnesium) during a thirty-minute period. The other 50 patients received a placebo.

This single dose of magnesium cut the incidence of postoperative dysrhythmias (dangerous abnormalities in heart rhythm) in half. There were eight dysrhythmias in the magnesium group and seventeen in the placebo group.

If we play the usual numbers game, we can say that this represents a 53-percent reduction in dysrhythmias. And the fraction of patients who benefited was 18 percent, not less than 1 percent, as we saw in the American aspirin study. In addition, these results could be seen in a sample size of 100 people, not thousands.

This study did have two weak features. First, only one dose of magnesium was given. The results could have been more impressive if the treatment group had continued to receive magnesium infusion as long as they were being fed intravenously (and even more if they had been instructed on how to continue supplementation orally after that). Second, the determination of magnesium levels was based on analyses of serum magnesium concentrations, which can give misleading readings after a serious trauma. Remember that trauma causes magnesium to leave muscle cells and migrate into the blood, so even though muscles may be seriously deficient in magnesium, the level of magnesium in the blood can be high, low, or normal. Amazingly, however, the data still showed positive results, even though the sample of patients was relatively small. If aspirin had been used in this study instead of magnesium, existing research reports indicate that no benefit would have been found.

People who use nutritional therapy have had the amount of plaque in their arteries decrease and their electrocardiograms restored to normal. Has this ever happened to someone taking aspirin? Has aspirin ever eliminated the tendency to have cardiovascular spasms or restored normal rhythm to the heart? Magnesium has!

Dr. Lloyd T. Iseri used magnesium therapy to restore life to a sample of four patients terminally ill from cardiac arrhythmia, congestive heart disease, and the toxic effects of an overdose of digoxin.[28] And magnesium restored me (a sample of one, admittedly) to normal health. My last attack of angina was more than four years ago, and I now have no limitations on my physical activities.

In reply to the question, "Which is better at reversing heart disease and preventing strokes, aspirin or magnesium?" magnesium is clearly the gold-medal winner.

Aspirin does not reduce arterial plaque; neither does it cause the development of collateral circulation in the form of newly generated blood vessels around partially blocked arteries. But we know that these benefits come to people who consume a healthful balance of essential vitamins and minerals, and refrain from eating harmful amounts of fats, oils, and proteins. These people also enjoy the antiplatelet actions of magnesium and vitamin E.

The Pritikin Clinics and Dr. Julian Whitaker's Wellness Institute use nutritional therapy to benefit patients on an individual basis.

If they had to depend on statistical averages to see a benefit in their patients (as aspirin therapy does), they would have to close their clinics. But their practices are flourishing, because individual patients are getting well. Their success rates are better than any reported for any of the conventional medical procedures used to treat people suffering from cardiovascular diseases. That includes even heart bypass surgery and coronary artery angioplasty—which, incidentally, have worse survival records than nutritional therapy does.

If you are one of the many people committed to a regimen of an aspirin a day, or every other day, you may wish to ask your physician to reconsider the treatment and explore the benefits of taking selected nutritional supplements and following a controlled diet. As a first step, I suggest that you—and your physician—read Dr. Julian M. Whitaker's book, *Reversing Heart Disease*.[29] A more recent specification of nutritional supplements and suggested dosages can be found in Dr. Richard A. Passwater's book, *The New Supernutrition*.[30] I also heartily recommend *The Doctor's Vitamin and Mineral Encyclopedia* by Dr. Sheldon Saul Hendler.[31]

When we put all of the facts together, we see that the aspirin phenomenon amounts to the widespread prescription of a familiar drug that seems to have, at best, a minimal beneficial effect in preventing cardiovascular disease. In fact, the data suggest that it may increase the risk of having a stroke.

Aspirin treatment may be somewhat beneficial to those who have already had a heart attack or stroke, but it seems to have little benefit for people who are trying to prevent a first seizure. And with the amazing benefits that can accrue from treating heart disease with nutritional methods, it seems absurd even to consider taking aspirin.

7

Organizing a Nutritional Therapy Plan and Making It Work

Once you decide you would like to try a nutritional approach to maintaining good health, the first thing to do is to consult with your physician about the advisability of changing your habits. Be prepared for the possibility that he or she may oppose the idea of taking mineral and vitamin supplements. A doctor is likely to tell you, "Vitamin and mineral supplements are unnecessary, since you get all of the minerals and vitamins you need by eating a balanced diet."

At this point, you can do one of a number of things. You can:

- Give in and say, "That sounds good to me." (Please *don't* do this.)
- Ask for a copy of such a balanced diet, along with specifications showing how it provides all of the essential nutrients you need each day. Insist on getting a list of the essential nutrients each food item contains. Then compare this with those presented in *Food Values of Portions Commonly Used.*[1] Remember that cooking food, exposing it to the air, or even storing it for a short time can destroy numerous sensitive nutrients. Consider that much of the food available in supermarkets is refined and so is deficient in nutrients.

- Ask what fraction of each nutrient from each food is actually absorbed from the digestive tract into the bloodstream. Nutrients that aren't absorbed don't do you any good.
- Remember that the recommended daily allowances in many cases are much lower than the amount of each nutrient you would need to support good health. This is true of magnesium and vitamins C and E, among others.
- Discuss the need to counteract the chemical free radicals we encounter in the environment. These include ozone from electric motors; nitrogen oxides from all combustion processes; chemicals present in chlorinated water; hydrocarbon chemicals and their derivatives present in automobile exhaust gases; the free radicals formed by excessive consumption of unsaturated fats (vegetable oils and partially hydrogenated vegetable oils); and many others. Ask for an explanation of the effects of these agents on your health.
- Even though your physician might deem them to be useless and a waste of money, ask if you can take selected nutritional supplements without harming your health. If the answer is yes, ask your doctor to help you select a program of essential vitamin and mineral supplements. Seek your doctor's help in monitoring the improvement in your health.
- If your physician denies that there is any point in taking nutritional supplements, lend him or her a copy of this book to read about the amazing health benefits of maintaining an adequate balance of minerals and vitamins in the body. Then discuss your personal case.
- If your doctor still claims that nutritional supplements are useless, or even harmful, ask for copies of original scientific studies that confirm that opinion.
- If you aren't satisfied that your doctor is taking your concerns seriously, you may wish to seek a second opinion, or even a third or fourth. If so, go ahead. It's your body and your health.

In any event, consult with your physician. It may not be easy, but be polite and persistent. You need his or her professional advice. And you may be surprised. Not all physicians resist nutritional therapies, though many do.

DEVELOPING A SUPPLEMENTATION PLAN

The principles set forth in this book should work for you without requiring you to make a major change in your eating habits. Merely adding a supplement program to your regular diet could work wonders. It did for me.

Unless you eat mostly fresh, uncooked or lightly cooked fruits and vegetables at least three or four times a day, you are very likely to be deficient in vitamins and minerals. Even if you do, you may end up with nutritional deficits. But don't expect to find a single supplement that will contain all of the nutrients you need, or have them in the right amounts.

Table 7.1 is a basic guide designed to help you structure your supplement program. It is organized with the mineral supplements listed first, followed by trace elements, and then the vitamins. The vitamins are subdivided into those that are predominantly oil or water soluble.

Deciding which supplements to take involves establishing priorities, exercising judgments, and making compromises. First, consider the priorities. Most people should start their list of supplements with magnesium, beta-carotene, vitamin A, vitamin C, vitamin E, and selenium. You can obtain vitamin A and selenium from a multivitamin product. Read the labels until you find one that lists both nutrients in acceptable concentrations. Each tablet should contain about 5,000 international units of vitamin A and 10 micrograms of selenium. Multivitamins should also reward you with all of the B vitamins and most of the trace elements you require.

Now we come to the judgments and compromises. Don't be concerned that the concentrations of a number of vitamins and minerals in your supplements are different from those listed in Table 7.1, and don't take more than the recommended doses listed on the product labels. If you are only beginning, you need time to get acquainted with the supplements and the way they affect you. You may need no more than the amount contained in a single tablet each day. With some supplements, particularly multivitamins, you may need no more than one tablet each week. Your requirements are unique. It is always best to err on the side of caution. Start with low dosages and work up to what you need.

Table 7.1 Suggested Daily Supplement Dosages

This table gives recommended amounts of nutrients for a daily supplement program, as well as a brief explanation of the chief functions of each. Note that suggested dosages for some nutrients are given in milligrams (mg), some in micrograms (mcg), and some in international units (IU). Both milligrams and micrograms are measures of weight, with 1 microgram equal to 1/1000 of 1 milligram. International units are measures of activity, not weight; the number of milligrams or micrograms in an international unit therefore varies, depending on the substance being measured.

Nutrient	Daily Dose	Function
MINERALS		
Calcium	800–1500 mg	Necessary for healthy bones, teeth, nerves.
Magnesium	200–1000 mg	Needed to maintain healthy nerve signals, muscle tone, cellular energy; helps prevent kidney stones; helps regulate blood pressure; helps to prevent heart disease.
Phosphorous	Supplements are probably unnecessary; phosphorus is abundant in foods. Check with your physician to learn your requirements.	Necessary for bones, teeth, cellular energy production, cell membranes.
Potassium	Enough to maintain potassium balance (you need magnesium to help the cells absorb potassium). Consult with your physician.	Helps regulate blood pressure, heart rhythm.
Zinc	15–35 mg	Assists immune system; helps heal wounds; helps cure colds; retards macular degeneration.

Nutrient	Daily Dose	Function
TRACE ELEMENTS		
Chromium	50–200 mcg	Increases glucose tolerance.
Copper	1.5–3 mg	Free radical scavenger.
Iodine (iodide salts)	150 mcg	Reverses hypothyroidism, goiter.
Iron	Less than 15 mg (avoid taking unless you have a deficiency; check with your physician)	Needed for red blood cells.
Manganese	2–5 mg	May be linked to brain health.
Molybdenum	75–250 mcg	Activates essential enzymes in the body.
Selenium	50–200 mcg	Stimulates immune system; protects against cancer; detoxifies heavy metals.
FAT-SOLUBLE VITAMINS		
Vitamin A	5,000 IU (1.5 mg)	Beneficial for eyes, skin, immune system; fights cancer.
Beta-carotene	25,000–100,000 IU (15–60 mg)	Used by the body to make vitamin A; destroys free radicals.
Vitamin D	Up to 400 IU (10 mcg)	Benefits immune system; protects against, treats cancer.
Vitamin E	400–800 IU (400–800 mg)	Antioxidant; protects against breast cancer, breast diseases, formation of blood clots; enhances immune response.

Nutrient	Daily Dose	Function
WATER-SOLUBLE VITAMINS		
Vitamin B_1 (Thiamin)	1–50 mg	Helps to counteract lead poisoning, heart disease, nerve disorders.
Vitamin B_2 (Riboflavin)	2–50 mg	Antioxidant.
Vitamin B_3 (Niacin)	20–100 mg (excessive doses should be avoided, as too much niacin can cause liver damage; consult with your physician)	Lowers cholesterol and risk of heart attack.
Vitamin B_5 (Pantothenic acid)	50–100 mg	Prevents, relieves rheumatoid arthritis; helps wounds heal.
Vitamin B_6 (Pyridoxine)	5–50 mg	Stimulates immune system and inhibits cancer.
Vitamin B_{12} (Cyanocobalamin)	5–50 mcg	Increases energy and sense of well-being; alleviates neuropsychiatric ailments; protects against cancer.
Biotin	100–300 mcg	Promotes healthy hair.
Choline	400–900 mg	Necessary for nerve signal transmission.
Folic acid	400 mcg	Protects against cancer and anemia; important for fetal development.
Vitamin C	2,000–10,000 mg (unless you have hemosiderosis or hemochromatosis; see page 119)	Stimulates immune system; helps prevent cancer; effective against colds.

Also, be aware that different amounts of nutrients affect the body in different ways. Deficiencies are bad for your health, but so are excesses. Don't make the mistake of thinking that more is always better. Figure 7.1 illustrates the way in which different dosages of essential nutrients affect the body's functions. Notice that death occurs at both the left (deficiency) and right (excess) intersections of the plot with the horizontal axis, which represents nutrient concentration. This illustrates the fact that while the body cannot live without an adequate amount of an essential nutrient, neither can it survive when the amount of that nutrient present goes beyond the toxic limit. Not only can you starve to death, you can also nourish yourself to death. The object in taking nutritional supplements is to get enough of all the essential nutrients to achieve good health, but not more.[2]

The ultimate indicator of the optimum amount of a particular tablet or nutrient will be how it affects you. Because each of us is a unique individual, nutritional requirements vary from person to person. You may wish to ask a pharmacist or a physician to help you select your supplements. Each person should use care to consume enough minerals and vitamins, but avoid taking an overdose. When you feel well, be cautious about increasing any dosage. Never take anything unless there is a specific need. These principles apply equally well to all vitamins, minerals, and other nutrients.

Let us turn now to a discussion of the different components of a sound supplementation plan.

MINERALS

The major metal ions in the body are calcium, magnesium, potassium, sodium, and zinc. All are necessary in the proper amounts for vibrant good health. If you choose to take only one mineral supplement, however, let it be magnesium. Chances are strong that you do not get enough of this vital nutrient from the food you eat.

Magnesium

One of the symptoms of a serious magnesium deficiency is an annoying condition of perpetual fatigue. The antidote is simply to include sufficient magnesium in your diet. You can do this by eating enough beans, peas, nuts, vegetables, and lean meat, and

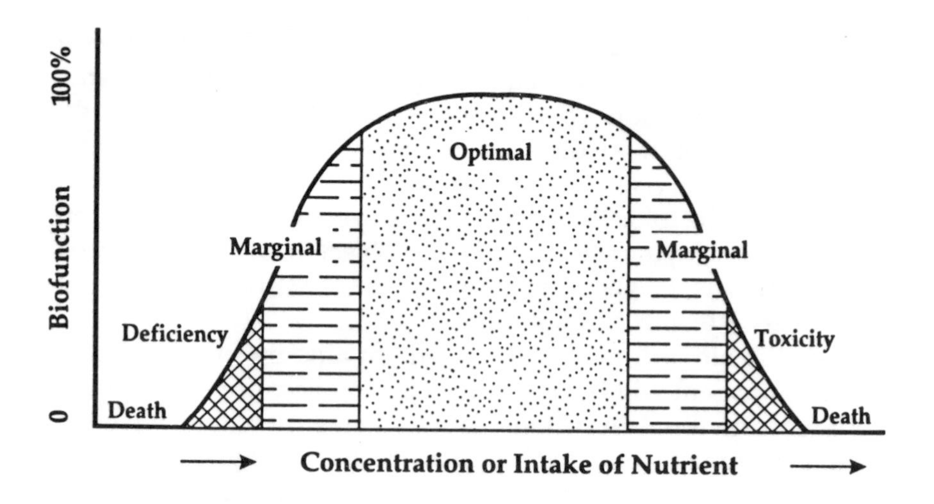

Figure 7.1 Effects of Different Amounts of Nutrients on the Body

The way in which different amounts of all essential nutrients in the body affect health can be visualized as a bell-shaped curve, with the horizontal axis representing the amount of the nutrient present and the vertical axis representing body function. At both the far left (deficiency) and right (excess) ends of the horizontal axis, body function is so impaired that death results. In between these two extremes are the toxic and deficient ranges at either end; the marginal ranges, in which the body is getting enough of the nutrient to support life (either just enough, or somewhat more than enough but not enough to be toxic); and the optimal range, in which the body receives the ideal amount it needs to be healthy. Keeping the amount of nutrients the body receives in the optimal range—but not beyond—is the goal of any good nutritional program.

Adapted from "The Essential Trace Elements," by Walter Mertz.[3]

by taking supplements in the form of magnesium salts or magnesium oxide. (You could take milk of magnesia [magnesium hydroxide], but magnesium tablets are more palatable and allow for better control over dosage.)

Depending on your requirements, you can take up to 1,000 milligrams (one gram) of magnesium per day in supplement form.

Some people may need more, but don't push the limits too far. You can get magnesium in tablet form at pharmacies and health food stores. It comes in tablets of different sizes and in a number of different compounds. You can also get magnesium in tablets that contain calcium as well. This choice takes care of the need for further supplementation with calcium.

If you need calcium as well as magnesium, you can take dolomite (calcium magnesium carbonate) or one of the other tablets that contains both elements. If you take dolomite, which is a natural form of limestone, you should ask about its purity. Some dolomites contain a small concentration of lead, which is within the legally allowable limits. However, lead accumulates in the body's tissues and can cause serious health problems, so it's best to avoid it.

Many people can benefit from taking both calcium and magnesium in tablets made from pure compounds of calcium and magnesium. These should contain no lead. Some supplements incorporate calcium and magnesium along with other nutrients, such as zinc or vitamin A. One convenient type of tablet contains magnesium chelated, or combined, with protein. Always read the labels carefully to learn how much *elemental* magnesium (or calcium, or any other mineral you wish to take) is contained in the tablet (more about this later in this chapter).

If you use a calcium-containing magnesium product exclusively, you will likely be taking a lot more calcium than you need. You may wish to take a tablet containing 500 milligrams of calcium and 250 milligrams of magnesium when you get up in the morning and another when you retire. During the day you can take one or two tablets containing 250 milligrams of magnesium without the calcium. Or you may wish to take these two elements separately, which gives you greater control over relative dosages. Whatever form of magnesium you take, spread out the doses as evenly as you can throughout the day.

On the other hand, people who consume unusually large amounts of calcium—such as cancer patients who take large quantities of shark cartilage—may need to take extra magnesium. It is possible that some individuals may develop excessive muscular tone as a result of consuming large amounts of calcium, which in turn could possibly lead to spasms and other consequences. Remember, calcium and magnesium are antagonists—they compete for chemical prominence in the cells. If you are taking shark cartilage, or are consuming large doses of calcium for some other reason, you may need to adjust your intake of supplemental magnesium accordingly. Consult with your physician.

The following is a reasonable approach to take when you begin taking magnesium supplements.

- Check with your physician to be sure you have no abnormal kidney function or other condition that would be complicated by taking magnesium supplements.
- Read the labels of a number of products. Become familiar with their formats.
- Select a product that contains either 100 or 250 milligrams of elemental magnesium (not total compound weight; see page 31) per tablet.
- You may wish to take calcium at the same time. Products containing both calcium and magnesium usually contain twice as much calcium as magnesium. That is perfectly all right.
- Start with a small dose, say one tablet per day for a week. A tablet containing 100 to 250 milligrams of magnesium is sufficient. Take your single tablet in the morning when you arise. Because an inordinate number of heart attacks occur around 9:00 A.M., you should significantly reduce your risk of heart attack if you take the tablet before 8:00 A.M.
- If you tolerate the magnesium tablets well for one week—and you should—you can take two tablets a day the second week. Take them about twelve hours apart. You should be able to take these tablets either with or without food. If you take them without food, the magnesium is likely to be absorbed more efficiently.
- If all is going well at the end of the second week, you may start taking a third tablet daily.
- If you are taking 250-milligram tablets, you should stay at the level of three tablets per day for a week or two before you increase the dosage to four tablets.

- If you think you would like to try another tablet, go ahead. But you probably will not need more than 1,000 milligrams per day on a regular basis. Let your body tell you when you achieve the optimum dosage.
- If you awaken at night and feel uptight, you may benefit by taking an extra magnesium tablet to relax your tense muscles. It's more healthful than taking sleeping pills.
- Once or twice on days when you become very angry or upset, you may wish to take a magnesium tablet to help keep your muscles from losing magnesium.

Finally, several words of caution. First, *do not confuse magnesium with manganese*. Read labels carefully. These elements are not the same, and they certainly do not perform the same functions in the body. Nor are they required in anywhere near equivalent amounts. It is unfortunate that these two elements have such similar names and that they are both furnished in tablet form as dietary supplements, because it can lead to confusion.

Second, *do not use antacid medications as a source of supplemental magnesium*. Some common antacids contain magnesium, but in a form that includes aluminum as a companion metal ion. Aluminum has been implicated as a possible contributor to Alzheimer's disease. Until more is known about this, do not expose yourself unnecessarily to that risk. Always read labels carefully.

Calcium

As people get older they need more calcium. This is due in part to a lower efficiency of absorption from the digestive tract. If you need to take a calcium supplement, up to 1 or 1.5 grams (1,000 to 1,500 milligrams) per day should be enough. Consult with your physician about this dosage. Also, the amount of calcium that will stay in your blood depends on your hormones and the balance of other electrolytes present. Ever since I have been taking measures to correct my magnesium deficiency, my blood calcium level has been low. The calcium concentration has remained consistently low, even though the supplemental calcium I took varied between 1 and 6 grams per day.

This underscores the fact that supplementation of calcium alone may not be enough to correct a deficiency. In the elderly, and in postmenopausal women in particular, other measures may be necessary. Consumption of a low-protein diet is one simple way to help the body retain and utilize calcium. This is true because the products of protein metabolism tend to react with calcium and carry it to the kidneys for excretion. To avoid this, if you are trying to correct a calcium deficiency, you should eat no more than two to three ounces of protein per day.

Phosphorus

This element is essential for giving strength to our bones and teeth. It is also a major component of adenosine triphosphate (ATP), a compound the body uses to extract energy from glucose. Phosphorus occurs abundantly in our foods. If you eat normally, you should not need phosphorus supplements.

Potassium

Potassium is necessary for the maintenance of normal heart rhythm and the health of the cardiovascular system. Check with your physician to determine your potassium requirements. Physicians often recognize potassium as one element that may need to be supplemented. Before you take any potassium supplements, you should have a blood analysis performed to determine whether or not you need it.

Unlike magnesium compounds, potassium salts are very soluble in water, and foods are about five to ten times richer in potassium than they are in magnesium. As a result, you may need to supplement magnesium but not potassium. Also, recall that no matter how much potassium is consumed, it cannot be replenished in the cells unless the magnesium status is adequate. If your potassium levels are too low, it may point to a deficiency of magnesium, rather than a need to consume more potassium.

Sodium

Because of the high content of sodium in most of our foods, you will not normally require supplements of that element. However, if you are elderly and taking a diuretic, or if your meals are not well balanced, consult with your physician to be assured that you do not have a sodium deficiency.

Zinc

Zinc, although a necessary mineral, should be taken with care. Some experts tell us to take no more than 35 milligrams per day. However, as you get older, you may need more. Zinc is beneficial in maintaining a healthy immune system. For men over fifty years of age, it can also help to relieve the adverse effects of an enlarged prostate gland—especially if zinc supplementation is combined with a low-fat diet. I personally take 100 milligrams of zinc per day.

TRACE ELEMENTS

The metallic minerals you need in trace quantities include iron, manganese, chromium, copper, and selenium, among others. For the most part, the recommended quantities of these elements should be present in your multivitamin tablets. Two of the trace elements, iron and selenium, deserve brief consideration on their own, however.

Iron should not be taken in supplement form unless your physician tells you that you need to do so. Too much iron in your system can lead to an excessive amount of toxic free radicals, and then to one or more of the serious degenerative diseases.

Selenium is an essential mineral needed in trace quantities. Many people decide to take it because it is an antioxidant and acts synergistically with vitamin E. However, it can become toxic if present in excessive amounts. You can get supplemental selenium over the counter in multivitamins and in tablets with dosages ranging from about 50 to 200 micrograms—*not* milligrams. (A microgram is 1,000 times smaller than a milligram.) If you wish to use selenium tablets, start with the lowest concentration you can buy and gradually increase the dosage. If you begin to feel drowsy and listless, or notice a metallic taste in your mouth, stop taking selenium immediately. When you feel better, you may or may not wish to try this supplement again at a lower dosage.

CHOOSING MINERAL SUPPLEMENTS

Metal ions (such as calcium, magnesium, potassium, and sodium) cannot occur alone. It is impossible to get a bottle of pure magnesium ions, for example. When you buy supplement tablets containing magnesium (or any other such ion), it will always be in combination with some other component or components that are ions bearing the opposite charge. These balance the ionic charge of the active ingredient, so that a stable compound is formed. The inert ingredients may also help the mineral to dissolve in the digestive juices. Always keep in mind that what you need to look at when you select your supplements is the content of the active element.

Some companies tend to write their labels in a way that seems to exaggerate the potency of their products. For example, I have a jar of magnesium tablets in which the magnesium is chelated, or combined, with amino acids. In large print the label says: MAGNESIUM 500 MILLIGRAMS. In smaller letters, under this eye-catching figure, it tells us that each tablet contains 100 milligrams of elemental magnesium. What this means is that, although each tablet weighs 500 milligrams, it actually contains only 100 milligrams of magnesium ion. This is the significant information for determining potency. As you choose your supplements, if you find that the information on a product label seems confusing, ask your pharmacist to help you interpret it. You can also refer to Table 2.1 to find the percentage of magnesium present in some of the more popular compounds (see page 32).

Another thing you must look for on product labels is the specification of dosage. Some labels tell you how much of each component is in one tablet, while others specify the amount contained in two or three tablets. Read the labels slowly and carefully, and be certain you understand exactly what each tablet contains.

Finally, many supplement labels state that the minerals the products contain are chelated. This merely means that the active components have been chemically combined with some other substance or substances. For example, to prevent calcium and magnesium tablets from forming a gritty powder, and to enhance their absorption in the digestive tract, some companies coat their tablets with protein and chelate (combine) the ions with amino acids or protein. Proteins and amino acids are natural materials compatible with the body, so they should have no ill effects. When these coatings and chelating agents perform their intended functions, they improve the quality of the product and may increase the absorption of the nutrients. (See Chelation on page 116.)

VITAMINS

Those of us who are concerned about nutrition are constantly faced with the nagging question of how we can know when we are deficient in a particular mineral or vitamin. Unfortunately, the standards set forth by the medical community are inadequate at best. For example, the recommended daily allowance (RDA) of vitamin C for an adult is 60 milligrams per day. That may be enough to keep a person from getting scurvy, but it is totally inadequate for maintaining a healthy body. I myself take 6,000 milligrams of vitamin C per day. That's 100 times the RDA!

Vitamin A and Beta-Carotene

The RDA for vitamin A is 5,000 international units, or 1.5 milligrams. Vitamin A deficiency degrades the immune system and causes night blindness and hard, scaly lesions on the skin. Vitamin A in large doses is toxic, however. The dose needed to cause toxicity varies with each individual, but most adults can probably take up to 15,000 international units of vitamin A per day with no ill effects. However, it is unnecessary to take vitamin A supplements at all if you take beta-carotene, because the body makes vitamin A from beta-carotene as it needs it, and beta-carotene does not have vitamin A's potential toxicity. The worst thing that will happen if you get too much beta-carotene is that your skin will start to turn a carroty yellow.

Like vitamin A, beta-carotene is usually measured in international units. Fortunately for the consumer, one international unit of beta-carotene is equivalent in activity to one international unit of vitamin A. (Because the compounds have different molecular

Chelation

Chelating agents are chemical species (molecules or ions) that form a molecular shield around a metal ion. This process generates a new, composite chemical species that has properties different from those of both the original metal ion and the chelating agent.

We may compare what happens in chelation to a man with a hang glider. As long as the hang glider is parked, both the man and the machine have their individual characteristics. However, when they take off and soar out together over the land below, they become a single unit that has the ability to do things that would have been impossible for either of them to have done alone.

The word "chelate" derives from the Latin word "chela," meaning claw. Like an eagle swooping down and grasping its prey in its claws, a chemical chelating agent or molecule grasps and surrounds a metal ion. The result is a new composite chemical species with properties different from those of either of the original components.

One common example of chelation may be found in the interaction of magnesium sulfate and water, which creates Epsom salts (known chemically as magnesium sulfate heptahydrate). Each formula unit of magnesium sulfate (represented by the chemical formula $MgSO_4$) is composed of one magnesium ion (the Mg^{2+}) and one sulfate ion (the SO_4^{2-}). When this compound is exposed to the air, it absorbs moisture. As illustrated below, six water molecules (H_2O) attach themselves symmetrically around each magnesium ion to form a spherical unit. They become attached by means of the oxygen (O) they contain (a seventh water molecule becomes attached to the sulfate ion, but for simplicity's sake these are omitted from the illustration).

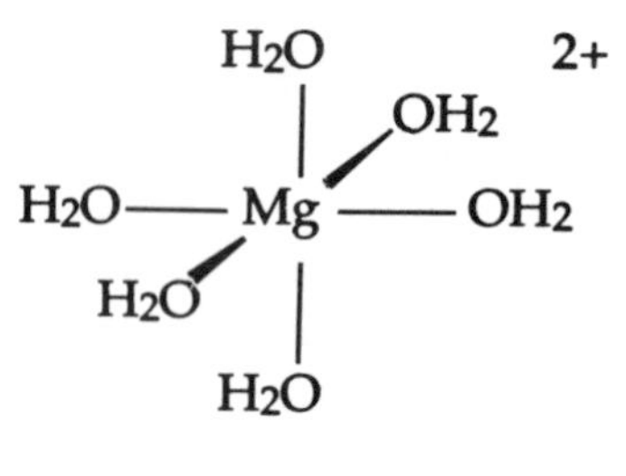

The attraction that causes the water and the magnesium to join together can easily be broken by heating the Epsom salts and driving off the water. When this is done, dry magnesium sulfate ($MgSO_4$) is re-formed. Thus the chelation of water with magnesium is a reversible process.

In the same way, the magnesium ions in our bodies interact with the oxygen and nitrogen atoms in the molecules of enzymes and other biochemical substances. These reactions cause muscles to contract, nerves to transmit their signals, and carbohydrates to yield their energy.

Not all chelation reactions are beneficial, however. For example, certain medications can interact with magnesium, calcium, iron, zinc, and other mineral ions in the body, causing the metal ions to become isolated and unable to perform their essential functions. They are then likely to be carried away and excreted, rather than used by the body. One common harmful chelation effect is the erosion of tooth enamel by ordinary table sugar. Sugar molecules contain an abundance of strategically located oxygen atoms that chelate the calcium ions in tooth enamel, causing it to gradually dissolve away. This is how eating sweet foods can lead to cavities and other dental problems.

Some chelation reactions can be either good or bad, depending on how and where they take place. For example, chewable vitamin C tablets can have the same effect as sugar on tooth enamel. I believe in supplementing my diet with vitamin C, but I am also the victim of severe tooth degeneration that was caused by using the chewable tablets. I became aware of this problem when my fillings started to fall out, one after another. (Vitamin C tablets designed for swallowing pose no such threat.) On the other hand, this chelating action of vitamin C probably contributes to its usefulness in dissolving unwanted calcium deposits in such places as the arteries *(see page 73)**.*

Minerals in dietary supplements are often chelated with protein, amino acids, or other materials to give the tablets more desirable properties, such as keeping them from being gritty. Chelation can also make a mineral easier for the body to absorb, or it can yield a product that gradually releases its active components over a period of time.

weights and there is an intervening chemical reaction, it takes about twice as much beta-carotene by weight as it does of vitamin A to yield the same number of international units.) It seems reasonable to suggest that most people should take between 25,000 and 100,000 international units (15 to 60 milligrams) of beta-carotene, and avoid taking any vitamin A supplements. If you take a multivitamin that contains vitamin A, however, that's all right because your body will convert beta-carotene into vitamin A only as it needs it. Both vitamin A and beta-carotene are antioxidants and tend to prevent the development of cancer.

The B Vitamins

A good multivitamin should furnish supplementation of your required B vitamins, along with the trace minerals you need. Read the labels and experiment with the products available. And don't buy a multivitamin solely because it is advertised to be just what a person in your age category needs. Some products that are targeted for senior citizens, for example, are much less potent (and at the same time often more expensive) than others that don't use that kind of advertising pressure. I usually try to get the multivitamin that has the most components in the highest concentrations. After you read a number of labels, you will begin to get a feeling for the appropriate concentration of each component. Spend some time at the vitamin counters in a number of pharmacies and health food stores and become familiar with the products that interest you.

Be especially alert to the content of the B vitamins. These include vitamins B_1 (thiamin), B_2 (riboflavin), B_3 (niacin), B_5 (pantothenic acid), B_6 (pyridoxine), and B_{12} (cyanocobalamin), as well as biotin, choline, folic acid, inositol, and para-aminobenzoic acid (PABA). You need the B vitamins on a daily basis because they are water soluble and the body doesn't store them. Of course, if you get all the B vitamins you need in your diet, it is unnecessary to take supplements. But as you get older, or if you are under physical or emotional stress, you will very likely benefit from taking supplements of these essential nutrients.

A deficiency in B vitamins can decrease mental acuity. In elderly people, this is sometimes misdiagnosed as senility. Vitamin B_{12}, folic acid, and choline are particularly important for mental function. Vitamins B_1, B_2, B_3, B_6, and inositol also contribute, in different ways, to the functioning of the brain and nervous system.

In the cells, vitamins B_1, B_2, and B_6, and biotin aid in the process of extracting energy from glucose. Folic acid, biotin, and pantothenic acid, together with vitamins B_6 and B_{12}, support and strengthen the immune system. The body neutralizes noxious chemicals through the action of vitamins B_2 and B_6. And both DNA and RNA—essential carriers of genetic information for every cell in the body—are the products of chemical processes that involve the active participation of vitamin B_{12} and folic acid.

Ask your physician about your requirements for these substances. You may discover that you don't need a whole multivitamin tablet each day. You may need only one-half or one-quarter of a tablet daily, or you may need just one tablet a week. Find out what works for you.

Vitamin C

For most people, there are distinct physical signs that can indicate when they are getting too little or too much vitamin C. If you bruise easily, you are probably deficient in vitamin C; if you take too much vitamin C, you will get diarrhea. There is usually a broad area of tolerance between these two extremes, but to maintain an optimum state of wellness you should try to keep your vitamin C level near the maximum. Most people should be able to tolerate at least two or three grams (2,000 to 3,000 milligrams) of vitamin C per day. Taking timed-release vitamin C tablets is a good way to ensure a relatively constant level of the vitamin in the bloodstream without the large surges in concentration that you get when you take conventional vitamin C tablets.

One note of caution: In people who have a genetic condition that causes iron overload (hemosiderosis or hemochromatosis), excess vitamin C can cause an abnormal uptake of iron, which can lead to kidney stones, diarrhea, irritation of the urinary tract, and heart disease. In such an individual, vitamin C supplementation can lead to serious illness, or even death. Before you begin to take vitamin C supplements, you should be certain that you do not belong to the approximately one-half of 1 percent of the population that suffers from these disorders.

Vitamin D

Vitamin D acts to increase the absorption and transport of calcium, and distributes calcium to the bones and teeth. There is some evidence suggesting an important role for vitamin D in the prevention and treatment of colorectal and breast cancer as well.

Exposure to specific frequencies of sunlight causes vitamin D to be synthesized in the skin. You can purchase a form of vitamin D (ergocalciferol) in supplement form, but be cautious—too much of it can be toxic. You should limit your supplements to about 400 international units per day. Excessive consumption of vitamin D causes calcification of soft tissue.

Vitamin E

Vitamin E emerged many years ago as an effective agent for preventing the formation of blood clots in the arteries, and even for helping to break up existing clots.[4] Unfortunately, most of those in the medical profession effectively ignored it, and continue to do so today. In my own experience, magnesium, vitamin E, and vitamin C form an unbeatable combination for maintaining the health of the heart and blood vessels.

Vitamin E is usually measured in international units. One international unit equals one milligram of d-alpha-tocopherol, the most active member of the vitamin E family of compounds. Vitamin E is comprised of seven different compounds called isomers. The most active of these are two substances called alpha-tocopherols, known as dextro- (d) and levo- (l), or d-alpha-tocopherol and l-alpha-tocopherol. A mixture of the two forms may be represented as dl-alpha-tocopherol; some supplement labels say only that the product contains mixed tocopherols. There are other tocopherols, but only the d-alpha and l-alpha forms are significantly active, and the structure known as d-alpha-tocopherol is the more active of the two.

Some products furnish vitamin E in a modified form. In these, the label will specify a name such as d-alpha-tocopheryl acetate. This means that the vitamin E entered into chemical reaction with acetic acid and formed a new chemical compound, without destroying its activity as a vitamin. The product is still active and acceptable.

How do you know when you need vitamin E? How much vitamin E should you take? There are no definitive signs of deficiency—until after you have developed some degenerative disease, such as atherosclerosis or cancer. Then you know that your immune system was not supported by adequate amounts of essential nutrients, especially vitamins A, C, and E, and beta-carotene.

The recommended daily allowance for vitamin E is 8 international units per day for women and 10 for men. *Ridiculously low* is too generous a description of these RDA values. In our environment, which is polluted with free radicals and other cancer-causing chemicals, most people should take at least 200 international units of vitamin E every day. The upper limit for vitamin E supplementation, based on the reading I have done, should be 800 international units per day. Until we gain valid information that supports the use of higher doses, we should therefore limit our consumption of vitamin E to that amount.

A FINAL WORD ABOUT SUPPLEMENTS

Many physicians will tell you that you are wasting your money if you buy nutritional supplements. Their advice usually is that if you eat a balanced diet, you are getting all of the nutrients you need to maintain good health. And you can consult tables of food values to learn the amounts of the different nutrients each food contains.

How wonderful life would be if it were that simple. Ideally, humans are designed to live off the land and eat grains, fruits, and vegetables that are freshly picked. And if you lived this way, you probably could get most of the nutrients you need from your diet. But we live in an industrialized era where processed foods, not fresh foods, are the norm. Storage, transportation, processing, and final food preparation all exact a serious toll from our food's nutritional quality.

During storage and transportation, food is in contact with the oxygen in the atmosphere. This alone destroys significant amounts of sensitive vitamins and other nutrients. Canning removes as much as 90 percent of sensitive nutrients; freezing destroys approximately 80 percent.

All of these processes purge from our foods the essential nutrients we need to keep us well. These include the antioxidants—vitamins A, C, and E, and beta-carotene. The B vitamins too fall victim to food processing. Even the mineral content of foods suffers from scalding and cooking. You might be surprised to learn that just cooking fresh vegetables can destroy over half of their vitamin content.

There are indications that many of us are indeed undernourished. Perhaps the most striking is that we have such high incidences of so many degenerative diseases—diseases not caused by viruses or germs. Among these are cardiovascular diseases, cancers, diabetes, macular degeneration, glaucoma, obesity, and arthritis. And as people increasingly base their diets on junk foods, we can only expect the situation to get worse. All of these factors make for a strong, valid argument in favor of using nutritional supplements.

THE IMPORTANCE OF DIET

It is possible to get tremendous benefits by adding nutritional supplements to your daily routine. But for a maximum, long-term improvement in your health, in addition to maintaining optimum levels of vital nutrients in your body, you should also convert from the typical modern diet, which is high in fats, meats, and refined foods, to one that relies more on complex carbohydrates, and less on protein and fat. First and foremost, calories from fats and oils should make up no more than 20 percent of your caloric intake. Carbohydrates, mostly complex carbohydrates (including whole grains, legumes, fruits, vegetables, breads, and cereals) should make up about 70 to 80 percent of your caloric consumption. You should limit proteins to approximately two or three ounces a day. This is sufficient for your body's needs and will take a load off of your kidneys.

Use the "rule of nine" to find the number of calories from fat in any food. Fats and vegetable oils contain approximately nine calories per gram, while carbohydrates and proteins have only four calories in each gram. So, for example, if a product label says that one serving contains 10 grams of fat (or vegetable oil, partially hydrogenated vegetable oil, etc.), multiply the 10 grams of fat by 9. The result—90—represents the number of calories in that one serving that are due to fat. If the food has a total of 300 calories per serving, that means that 30 percent of its calories come from fat. (Divide the 90 calories from fat by the total number of calories, 300. This gives you 0.30. Then multiply that number by 100 to get it expressed as a percentage.) Since that exceeds your limit of 20 percent of dietary calories from fats, you would either avoid buying the product or eat it with some other food that has a much lower fat content.

Similarly, you can find out the fraction of calories due to either carbohydrates or proteins by taking the number of grams of either and multiplying by four (instead of nine). Then do the rest of the calculation just as you would for fats and oils.

Regular exercise and a good diet that limits fats will help you maintain optimal health in virtually every respect. Eat red meat, poultry, and seafood sparingly, if at all. Minimize your use of foods that contain milk, cheese, or butter. Avoid *all* foods that contain partially hydrogenated vegetable oils. When you must use oils, use only pure vegetable oils, and use as little of them as you can. Read the labels on all packaged foods carefully.

Adhering to such a diet will take away the fun of dining on a twelve-ounce steak or prime rib. But changing your diet will likely give you a lot more time to enjoy life in other ways, as your health improves and your longevity increases. You will probably feel so well that you will want to exercise. Please do, and you will feel even better. (Just be careful to avoid activities that exceed your capabilities. Consult with your physician.)

Your efforts will reward you with freedom from fear of a heart attack or stroke; reversal of cholesterol-plaque formation; and a much lower risk of developing such common, progressively debilitating diseases as diabetes, cancer, acute macular degeneration, cataracts, osteoporosis, arthritis, gout, and kidney failure. As a result, you will also feel an enhanced sense of well-being and have a positive outlook on life.

In your low-fat diet, be sure to include many healthful foods that contain magnesium. Our bodies can't make magnesium, but we can get it by eating beans, grains, nuts, and fruits. Soybeans are especially rich in magnesium. Because every molecule of chlorophyll contains one atom of magnesium at its center, all green vegetables are sources of magnesium as well. A list of the magnesium, potassium, and phosphorous contents of selected foods appears in Table 7.2.

Of the foods we most commonly eat, nuts and seeds are the richest in magnesium, followed by the legumes (beans and peas).[5] The cartoon character Popeye, with his fondness for spinach, can teach us something about nutrition. Spinach has a magnesium content close to that of soybeans. Other vegetables and meats have significant, although somewhat lower, levels of magnesium. Among fresh fruits, bananas furnish the most magnesium and potassium. Raisins and dates, because they are dried, have an enhanced percentage of both nutrients.

In their packaged forms, some specialty foods furnish exceptionally high concentrations of magnesium, potassium, and phosphorus. These include dehydrated seaweed, soybean protein concentrate, and tomato powder. However, it is unlikely that anyone would consume any of these in their original packaged form, so you will dilute their food values as you prepare them to be eaten. Chocolate also has a high magnesium content, but consumption of large quantities of chocolate would be unwise, and cannot be recommended.

If you eat well-balanced meals, the amounts of potassium and phosphorus you get from food will probably be adequate to supply what you need. Potassium and magnesium together help regulate the heart's rhythm, and both phosphorus and magnesium are essential in generating the energy for muscular motion. The recommended daily allowance for phosphorus is 800 milligrams for all adults, with an additional 400 milligrams recommended for pregnant and lactating women. For potassium, the RDA is not well established, and because there are inherent dangers in potassium supplementation for people who are taking diuretic and other medications, I suggest that you consult with your physician and determine from the results of blood tests what your potassium requirements really are.

That you should supplement your diet with magnesium and not

Table 7.2 Mineral Content of Selected Foods

The foods listed in this table are good sources of the essential minerals magnesium, potassium, and phosphorus. The quantities of each mineral are expressed as milligrams of mineral per 100 grams (approximately 3½ ounces) of food item.

Food	Magnesium	Potassium	Phosphorus
FRUITS			
Apple, fresh (uncooked)	4	115	7
Banana, fresh (uncooked)	29	396	19
Cantaloupe, fresh (uncooked)	11	309	17
Dates, dried	35	652	40
Orange, navel, fresh (uncooked)	11	179	19
Peach, fresh (uncooked)	7	197	13
Pineapple, fresh (uncooked)	14	113	13
Plum, fresh (uncooked)	6	171	11
Raisins, seedless	35	746	115
Raspberries, fresh (uncooked)	18	152	12
Strawberries, fresh (uncooked)	11	167	19
MEATS			
Bacon, cured (pan-fried)	26	484	337
Chicken, without skin (fried)			
dark meat	25	253	187
white meat	29	263	231
Turkey, without skin (roasted)			
dark meat	24	290	204
white meat	25	225	271
Beef, ground (broiled)			
medium	21	313	161
well done	25	369	190
Beef heart (simmered)	25	133	250
Pork or ham (roasted)	22	409	281
Sausage, fresh (cooked)	18	459	185
MILK			
Dry, whole	84	1331	775
Fresh			
3.3 percent fat	14	152	93
1 percent fat	14	156	96

Food	Magnesium	Potassium	Phosphorus
NUTS AND SEEDS			
Almonds (dry roasted)	307	782	557
Cashews (dry roasted)	264	571	496
Peanuts (dry roasted)	175	657	357
Pumpkin seeds (roasted)	543	818	1189
VEGETABLES			
Asparagus (boiled)	19	310	60
Avocado (raw)	40	634	42
Beans			
green (boiled)	25	298	39
kidney (boiled)	47	417	147
lima (boiled)	44	508	111
navy (boiled)	59	368	157
Beets (boiled)	36	313	31
Broccoli (raw)	25	168	34
Brussels sprouts, frozen (cooked)	24	326	54
Corn (boiled)	32	249	102
Cowpeas (boiled)	53	278	39
Peas, green (raw)	21	543	46
Seaweed, dried	770	1125	52
Soybeans, mature (boiled)	102	515	244
Spinach (raw)	79	557	50
Squash, summer (raw)	23	194	35
Sweet potato (baked)	28	348	54
MISCELLANEOUS			
Baking chocolate	304	832	418
Soy protein concentrate	314	2204	839
Tomato powder	178	1927	295

Adapted from Bowes and Church's *Food Values of Portions Commonly Used*. 15th edition, revised by Jean A.T. Pennington.[6]

with potassium may sound contradictory. However, on average, the content of potassium in foods is about five to ten times greater than that of magnesium. Also, compounds of potassium are more soluble in the body's fluids than those of magnesium are. Consequently, potassium is more readily absorbed into the bloodstream.

And again we must consider that all-important factor—that potassium will not be utilized by the cells unless they contain an adequate level of magnesium.[7]

I suggest that you consider using both a low-fat diet and a regimen of food supplements to maintain your electrolyte balance and meet your vitamin requirements. In particular, you should consume sufficient quantities of the antioxidants beta-carotene, vitamin C, and vitamin E to control excess free radicals.

THE IMPORTANCE OF A HEALTHY ATTITUDE

The third major element in your quest for optimum health should be confronting the stress in your life—or more precisely, the way in which you deal with the stresses in your life. Stress initiates biochemical processes that deplete the body of essential electrolytes and vitamins, and lays the foundation for such ailments as high blood pressure, muscular and arterial spasms, and cardiac arrhythmia, in addition to mental anguish. This is particularly true of chronic stress that comes from worrying about the outcome of situations over which you have no control.

Like most people in modern society, I know from experience what it is like to be in essentially hopeless situations, and I must confess that I was highly proficient at worrying about them. I have no doubt that this was one of the causes of my illness. And my resolve to overcome it helped lead to my recovery.

We often worry needlessly about things over which we have no control. My suggestion is, if you can't change something, don't worry about it. Accept it, and seek ways to get what you can out of it—or at least avoid as much harm as possible.

Focus on doing the best you can without compromising your principles. You may not like your situation, but if you try to oppose stronger, immovable forces, you will surely lose. I clearly remember an oppressive situation to which I was continuously subjected and which I could not improve. I worried and fretted and became ill. When I went to see my cardiologist, he inquired about the things that were bothering me. And when I enumerated the problems I considered to be so stressful, he asked, "Who are you hurting [by worrying about them]?" Facing that question was one of the milestones in my recovery.

Much stress in life arises from what some of us call "people problems." People tend to irritate us, especially if they look, think, or act differently than we do. But this is the real world, and each of us has a place in it. Instead of becoming irritated, try to view personality differences as interesting contrasts. It may help if you focus on using your observations of differences to predict how certain stimuli will affect the individuals you are observing. (But keep these thoughts rigorously to yourself. Don't even think of revealing them to anyone. Just play the little game alone, a sort of personality solitaire.)

My next suggestion for beating stress is the most important, and often the hardest to implement: Learn to forgive. Learn to forgive people absolutely and completely, without hanging on to any memory of the offense. It's all right to file away memories that may help you defend yourself in certain situations later, but don't take offense. Mentally step back and view the situation from a detached point of view. Hold no grudges. Grudges only serve to poison your mind and generate stress.

Also, it is of paramount importance not to engage in gossip. Not only will you find this makes you feel much better, but your coworkers and associates will be more comfortable in your presence. Expend your efforts instead on constructive thoughts and activities. This will surely get the attention of those around you, and result in an environment with an enhanced potential for health, happiness, and advancement in every enterprise.

You are now ready to begin enjoying a state of health accessible only to those who are well nourished, both physically and spiritually. For starters, you should begin a supplement program that suits you. At the same time, you can start increasing your consumption of healthy foods and cutting back on your intake of fats and oils.

As you begin to feel better and view the world through healthier and more peaceful eyes, you can start reducing your stress by ceasing to be offended by the behavior and attitudes of others. Chart your own course in whatever situation you find yourself; act independently of the attitudes of others. Just be calm and pleasant, and play your little private game of personality solitaire. It may be difficult at first, but as time passes, you will begin to see people in more gracious shades of light. And remember, the person who will benefit most from this attitude will be you.

When you have gained confidence in your program of improvement, you can really start to modify your diet. Read the labels on all packaged foods. Reject those that contain harmful substances such as partially hydrogenated vegetable oils, which are prone to form unnatural chemical free radicals in your body. This act alone will probably send you scurrying to the local health food stores to find acceptable brands of such products as crackers and pastries.

While you're in the health food store, you will find books that will help you learn how to prepare grains and vegetables in the most healthful ways. At first, you may think these foods taste somewhat bland, but stay with your dietary resolve for several weeks. By that time you will have learned to appreciate their more subtle tastes—in fact, you will probably enjoy them more than the food you used to eat. Steam or lightly cook your vegetables for one to ten minutes to minimize the loss of nutrients, and you'll find that they are actually tastier when they are still crisp. Try to avoid adding salt to your foods. You'll grow to find their natural taste more pleasant, and food is more healthful without salt.

This program does require the consistent application of discipline. There will probably be times when you will be tempted to return to your previous lifestyle. When that happens, reflect on the improvement you see in your health, and keep working to adhere faithfully to your new, healthful way of life. Your reward will be a longer, richer life that is filled with more joy and peace of mind than you have ever experienced before. This works for me, and it can work for you, too.

Notes

Introduction
It Happens Without Warning

1. James Landauer, "Aspirin and Heart Attacks—A Flawed Study?" *Privileged Information*, 1 June 1988.
2. Steering Committee of the Physicians' Health Study Research Group, "Preliminary Report: Findings From the Aspirin Component of the Ongoing Physicians' Health Study," *The New England Journal of Medicine* 318(4) (1988), 262–264.

   Steering Committee of the Physicians' Health Study Research Group, "Final Report on the Aspirin Component of the Ongoing Physicians' Health Study," *The New England Journal of Medicine* 321(3) (1989), 129–135.
3. Richard Peto et al., "Randomized Trial of Prophylactic Daily Aspirin in British Male Doctors," *British Journal of Medicine* 296 (1988), 313–316.
4. James Landauer, "Aspirin and Heart Attacks—A Flawed Study?"
5. Bo K. Siesjö, "Calcium and Cell Death," *Magnesium* 8 (1989), 223–237.

6. Bert Stern et al., *The Pill Book*, 4th ed. (New York: Bantam Books, 1990).
7. R.S. Parsons, T.C. Butler, and E.P. Sellars, "Hardness of Local Water Supplies and Mortality From Cardiovascular Disease," *Lancet* iii (1961), 213.

Chapter 1
Magnesium's Role in Our Lives

1. Stanley E. Manahan, *Environmental Chemistry* (Boston: Willard Grant Press, 1979).
2. J.K. Aikawa, *Magnesium: Its Biological Significance* (Boca Raton, FL: CRC Press, Inc., 1981).
   J.W. Schopf, "The Evolution of the Earliest Cells," *Scientific American* 239 (1978), 110–140.
3. J.W. Schopf, "The Evolution of the Earliest Cells."
4. Sheldon Saul Hendler, *The Doctor's Vitamin and Mineral Encyclopedia* (New York: Simon and Schuster, 1990), 496.
5. Bo K. Siesjö, "Calcium and Cell Death," *Magnesium* 8 (1989), 223–237.
6. Steering Committee of the Physicians' Health Study Research Group, "Preliminary Report: Findings From the Aspirin Component of the Ongoing Physicians' Health Study," *The New England Journal of Medicine* 318(4) (1988), 262–264.
   Steering Committee of the Physicians' Health Study Research Group, "Final Report on the Aspirin Component of the Ongoing Physicians' Health Study," *The New England Journal of Medicine* 321(3) (1989), 129–135.

Chapter 2
Why We Have Magnesium Deficiencies

1. J.R. Marier and L.C. Neri, "Quantifying the Role of Magnesium in the Interrelationship Between Human Mortality/Morbidity and Water Hardness," *Magnesium* 4 (1985), 53–59.
   J.R. Marier, "Magnesium Content of the Food Supply in the Modern-Day World," *Magnesium* 5 (1986), 1–8.

A. Roblin, "New Clues to the Power of Magnesium," *Prevention*, April 1989, 33–39.

2. Heljä Pitkänen, "Industrial Possibilities to Interfere With the Salt Problem: Dietary Na/(K+Mg) Ratio," *Magnesium* 1 (1982), 298–303.

3. Ibid.

4. R.S. Parsons, T.C. Butler, and E.P. Sellars, "Hardness of Local Water Supplies and Mortality From Cardiovascular Disease," *Lancet* iii (1961), 213.

   J.R. Marier and L.C. Neri, "Quantifying the Role of Magnesium in the Interrelationship Between Human Mortality/Morbidity and Water Hardness."

   J.R. Marier, "Magnesium Content of the Food Supply in the Modern-Day World."

   H.A. Schroeder, "Municipal Drinking Water and Cardiovascular Death Rates," *Journal of the American Medical Association* 195(2) (1966), 81/125–85/129.

   H.A. Schroeder, "Relation Between Mortality From Cardiovascular Disease and Treated Water Supplies," *Journal of the American Medical Association* 172 (1960), 98/1902–104/1908.

   T.W. Anderson et al., "Ischemic Heart Disease, Water Hardness and Myocardial Magnesium," *Canadian Medical Association Journal* 113 (1975), 199–203.

   J.R. Marier, "The Role of Environmental Magnesium in Cardiovascular Diseases," *Magnesium* 1 (1982), 266–276.

   J. Durlach et al., "Magnesium Levels in Drinking Water and Cardiovascular Risk Factor: A Hypothesis," *Magnesium* 4 (1985), 5–15.

5. R.S. Parsons, T.C. Butler, and E.P. Sellars, "Hardness of Local Water Supplies and Mortality From Cardiovascular Disease."

6. A. Roblin, "New Clues to the Power of Magnesium."

7. W.J. MacLennan, "Diuretics in the Elderly: How Safe?" *British Medical Journal* 296 (1987), 1551.

8. Lloyd T. Iseri, B.J. Allen, and M.A. Brodsky, "Magnesium Therapy of Cardiac Arrhythmias in Critical-Care Medicine," *Magnesium* 8 (1989), 299–306.

9. T. Regan and P. Ettinger, "Varied Cardiac Abnormalities in Alcoholics," *Alcoholism* 3 (1979), 40–49.

10. K.B. Franz, "Magnesium Intake During Pregnancy," *Magnesium* 6 (1987), 18–27.

Guillermo J. Valenzuela and Laurel A. Munson, "Magnesium and Pregnancy," *Magnesium* 6 (1987), 128–135.

11. Lloyd T. Iseri, B.J. Allen, and M.A. Brodsky, "Magnesium Therapy of Cardiac Arrhythmias in Critical-Care Medicine."
12. Leo D. Galland, "Magnesium in Neuropsychiatric Disorders," *Magnesium* 9 (1990), 324.
13. J. McL. Howard et al., "Magnesium and Chronic Fatigue Syndrome," *Lancet* 340 (1992), 426.
14. K.B. Franz, "Magnesium Intake During Pregnancy."
15. Hans-Georg Classen, "Magnesium and Potassium Deprivation and Supplementation in Animals and Man: Aspects in View of Intestinal Absorption," *Magnesium* 3 (1984), 257–264.
16. T. Regan and P. Ettinger, "Varied Cardiac Abnormalities in Alcoholics."
17. Hans-Georg Classen, "Magnesium and Potassium Deprivation and Supplementation in Animals and Man: Aspects in View of Intestinal Absorption."
18. Mildred S. Seelig, "Magnesium Requirements in Human Nutrition," *Magnesium Bulletin* 3 (1981), 26–27.
19. Ibid.
20. C.S. Anast and D.W. Gardner, "Magnesium Metabolism," in Coburn Bronner, *Disorders of Mineral Metabolism*, Vol. III (New York: Academic Press, 1981), 423–506.
21. Hans-Georg Classen, "Magnesium and Potassium Deprivation and Supplementation in Animals and Man: Aspects in View of Intestinal Absorption."
22. Julian Whitaker, *Health and Healing*, Vol. 1, No. 2 (September 1991), 3.

Chapter 3
A Warning of Heart Disease: Hypertension

1. Stevo Julius et al., "The Association of Borderline Hypertension With Target Organ Changes and Higher Coronary Risk,"

Journal of the American Medical Association 264(3) (1990), 354–358.

2. Adam Timmis, "Modern Treatment of Heart Failure," *British Medical Journal* 297 (1988), 83.
3. W.J. MacLennan, "Diuretics in the Elderly: How Safe?" *British Medical Journal* 296 (1987), 1551.
4. Theodore D. Mountokalakis, "Diuretic-Induced Magnesium Deficiency," *Magnesium* 2 (1983), 57.
5. Burton M. Altura and Bella T. Altura, "New Perspectives on the Role of Magnesium in the Pathophysiology of the Cardiovascular System, II. Experimental Aspects," *Magnesium* 4 (1985), 245–271.
6. Bo K. Siesjö, "Calcium and Cell Death," *Magnesium* 8 (1989), 223–237.
7. Maurice E. Shils, "Experimental Production of Magnesium Deficiency in Man," *Annals of the New York Academy of Sciences* 162 (1969), 847–855.
8. Ibid.
9. Maurice E. Shils, "Experimental Production of Magnesium Deficiency in Man."

   Michael P. Ryan, Robert Whang, and W. Yamalis, "Effect of Magnesium Deficiency on Cardiac and Skeletal Muscle Potassium During Dietary Potassium Restriction," *Proceedings of the Society for Experimental Biological Medicine* (1973), 1045–1047.

   Robert Whang and J.K. Aikawa, "Magnesium Deficiency and Refractoriness to Potassium Repletion," *Journal of Chronic Diseases* 30 (1977), 65–68.

   Robert Whang, "Magnesium and Potassium Interrelationships in Cardiac Arrhythmias," *Magnesium* 5 (1986), 127–133.
10. Maurice E. Shils, "Experimental Production of Magnesium Deficiency in Man."
11. Michael P. Ryan, Robert Whang, and W. Yamalis, "Effect of Magnesium Deficiency on Cardiac and Skeletal Muscle Potassium During Dietary Potassium Restriction."
12. J.G. Henrotte, "Type A Behavior and Magnesium Metabolism," *Magnesium* 5 (1986), 201–210.

13. Luther Clark, "Role of Electrolytes in the Etiology of Alcohol-Induced Hypertension," *Magnesium* 8 (1989), 124–131.
14. Ibid.
15. Burton M. Altura and Bella T. Altura, *Clin. Exp. Res.* 11 (1987), 99–111.
16. Bernard Horn, "Magnesium and the Cardiovascular System," *Magnesium* 6 (1987), 109–111.

Chapter 4
Cardiovascular Disease

1. Dean Ornish, *Dr. Dean Ornish's Program for Reversing Heart Disease* (New York: Random House, 1990), *xxiv*.
2. L. Cohen and R. Kitzes, "Prompt Termination and/or Prevention of Cold-Pressor-Stimulus-Induced Vasoconstriction of Different Vascular Beds by Magnesium Sulfate in Patients With Prinzmetal's Angina," *Magnesium* 5 (1986), 144–149.
3. Ibid.
4. L. Cohen and R. Kitzes, "Prompt Termination and/or Prevention of Cold-Pressor-Stimilus-Induced Vasoconstriction of Different Vascular Beds by Magnesium Sulfate in Patients With Prinzmetal's Angina."

   Myron Prinzmetal et al., "Angina Pectoris I. A Variant Form of Angina Pectoris," *American Journal of Medicine* 27 (1959), 375–388.
5. A. Flanagin, "JAMA 100 Years Ago: Raynaud's Disease by G.M. Garland, M.D., of Boston," *Journal of the American Medical Association* 262 (1989), 3290.
6. Myron Prinzmetal et al., "Angina Pectoris I. A Variant Form of Angina Pectoris."
7. J.S. DeCaestecker, "The Oesophagus as a Cause of Recurrent Chest Pain: Which Patients Should Be Investigated and Which Tests Should Be Used?" *Lancet* (1985), 1143–1146.

   R.W. Dhurandhr et al., "Prinzmetal's Variant Form of Angina With Arteriographic Evidence of Coronary Arterial Spasm," *American Journal of Cardiology* 30 (1972), 9002–9905.
8. C.B. Higgins et al., "Clinical and Arteriographic Features of

Prinzmetal's Variant Angina: Documentation of Etiologic Factors," *American Journal of Cardiology* 37 (1976), 831–839.

9. Burton M. Altura, "Ischemic Heart Disease and Magnesium," *Magnesium* 7 (1988), 57–67.
10. Myron Prinzmetal et al., "Angina Pectoris I. A Variant Form of Angina Pectoris."
11. Isabelle Joris and Guido Majno, "Endothelial Changes Induced by Arterial Spasm," *American Journal of Pathology* 102 (1981), 346–358.

    S. David Gertz et al., "Endothelial Cell Damage and Thrombus Formation After Partial Arterial Constriction: Relevance to the Role of Coronary Artery Spasm in the Pathogenesis of Myocardial Infarction," *Circulation* 63 No. 3 (1981), 476–486.
12. Isabelle Joris and Guido Majno, "Endothelial Changes Induced by Arterial Spasm."
13. Ibid.
14. S. David Gertz et al., "Endothelial Cell Damage and Thrombus Formation After Partial Arterial Constriction: Relevance to the Role of Coronary Artery Spasm in the Pathogenesis of Myocardial Infarction."
15. S. David Gertz et al., "Effect of Magnesium Sulfate on Thrombus Formation Following Partial Arterial Constriction: Implications for Coronary Vasospasm," *Magnesium* 6 (1987), 225–235.
16. C. Ezzell, "Cell Channel Finders Garner Medical Nobel," *Science News* 140 No. 15 (12 October 1991), 231.

    R. Dagani, "Ion Channels: Discoverers Win Physiology Nobel," *Chemical and Engineering News* 69 No. 41 (14 October 1991) 4, 5.
17. Ibid.
18. Bert Stern et al., *The Pill Book*, 4th ed. (New York: Bantam Books, 1990), 266, 594–595, 917.
19. Robert Whang, "Magnesium and Potassium Interrelationships in Cardiac Arrhythmias," *Magnesium* 5 (1986), 127–133.
20. Michael P. Ryan, Robert Whang, and W. Yamalis, "Effect of Magnesium Deficiency on Cardiac and Skeletal Muscle Potas-

sium During Dietary Potassium Restriction," *Proceedings of the Society for Experimental Biological Medicine* (1973), 1045–1047.

21. Lloyd T. Iseri, "Magnesium and Cardiac Arrhythmias," *Magnesium* 5 (1986), 111–126.
    E.B. Flink, "Magnesium Deficiency—Etiology and Clinical Spectrum, *Acta Med. Scand. Suppl.* 647 (1981), 125–137.
22. R.H. Helfant, "Hypokalemia and Arrhythmias," *American Journal of Medicine* 80, supplement 4A (1986) 13–22.
23. Robert Whang, "Magnesium and Potassium Interrelationships in Cardiac Arrhythmias."
24. Lloyd T. Iseri, "Magnesium and Cardiac Arrhythmias."
25. Milton Packer, Stephen S. Gottlieb, and Paul D. Kessler, "Hormone-Electrolyte Interactions in the Pathogenesis of Lethal Cardiac Arrhythmias in Patients With Congestive Heart Failure," *American Journal of Medicine* 80, supplement 4a (1986), 23–29.
26. Milton Packer, Stephen S. Gottlieb, and Paul D. Kessler, "Hormone-Electrolyte Interactions in the Pathogenesis of Lethal Cardiac Arrhythmias in Patients With Congestive Heart Failure.
    Burton M. Altura and Bella T. Altura, "Biochemistry and Pathophysiology of Congestive Heart Failure: Is There a Role for Magnesium?" *Magnesium* 5 (1986), 134–143.
27. Bert Stern et al., *The Pill Book,* 366–368.
28. Ibid, 547, 575–576, 708, 749, 853–854.
29. Milton Packer, Stephen S. Gottlieb, and Paul D. Kessler, "Hormone-Electrolyte Interactions in the Pathogenesis of Lethal Cardiac Arrhythmias in Patients With Congestive Heart Failure."
    Burton M. Altura and Bella T. Altura, "Biochemistry and Pathophysiology of Congestive Heart Failure: Is There a Role for Magnesium?"
    Wai H. Lee and Milton Packer, "Prognostic Importance of Serum Sodium Concentration and Its Modification by Converting-Enzyme Inhibition in Patients With Severe Chronic Heart Failure," *Circulation* 73 (1986), 257–267.
    Milton Packer, Stephen S. Gottlieb, and Mark A. Blum,

"Immediate and Long-Term Pathophysiologic Mechanisms Underlying the Genesis of Sudden Cardiac Death in Patients With Congestive Heart Failure," *American Journal of Medicine* 82, supplement 3A (1967), 4–10.

30. M.J. Brown, D.C. Brown, and M.B. Murphy, "Hypokalemia From Beta 2-Receptor Stimulation by Circulating Epinephrine," *The New England Journal of Medicine* 309 (1983), 1414–1419.

    A.D. Struthers, R. Whitesmith, and J.L. Reid, "Prior Thiazide Diuretic Treatment Increases Adrenaline-Induced Hypokalemia," *Lancet* (1983), 1358–1361.

31. Milton Packer, Stephen S. Gottlieb, and Paul D. Kessler, "Hormone-Electrolyte Interactions in the Pathogenesis of Lethal Cardiac Arrhythmias in Patients With Congestive Heart Failure."

    Burton M. Altura and Bella T. Altura, "Biochemistry and Pathophysiology of Congesive Heart Failure: Is There a Role for Magnesium?"

    Wai H. Lee and Milton Packer, "Prognostic Importance of Serum Sodium Concentration and Its Modification by Converting-Enzyme Inhibition in Patients With Severe Chronic Heart Failure."

    Milton Packer, Stephen S. Gottlieb, and Mark A. Blum, "Immediate and Long-Term Pathophysiologic Mechanisms Underlying the Genesis of Sudden Cardiac Death in Patients With Congestive Heart Failure."

32. Lloyd T. Iseri, B.J. Allen, and M.A. Brodsky, "Magnesium Therapy of Cardiac Arrhythmias in Critical-Care Medicine," *Magnesium* 8 (1989), 299–306.

33. I.H. Chaudry et al., "Use of Magnesium-ATP Following Liver Ischemia," *Magnesium* 7 (1988), 68–77.

34. I.H. Chaudry, M.G. Clemens, and A.E. Baue, "Alterations in Cell Function With Ischemia and Shock and Their Correction," *Archives of Surgery* 116 (1981), 1309–1317.

35. Milton Packer, Stephen S. Gottlieb, and Paul D. Kessler, "Hormone-Electrolyte Interactions in the Pathogenesis of Lethal Cardiac Arrhythmias in Patients With Congestive Heart Failure."

36. Ibid.

Chapter 5
Heart Attack and Stroke

1. U.P. Steinbrecher et al., "Modification of Low Density Lipoprotein by Endothelial Cells Involves Lipid Peroxidation and Degradation of Low Density Lipoprotein Phospholipids," *Proceedings of the National Academy of Sciences of the United States of America* 83 (1984), 3883–3887.

   D.W. Morel, P.E. DiCorleto, and G.M. Chisolm, "Endothelial and Smooth Muscle Cells Alter Low Density Lipoprotein in Vitro by Free Radical Oxidation," *Arteriosclerosis* 4 (1984), 357–364.

2. Daniel Steinberg et al., "Beyond Cholesterol: Modifications of Low-Density Lipoprotein That Increase Its Atherogenicity," *The New England Journal of Medicine* 320 (1989), 915–924.

3. Matthias Rath and Linus Pauling, "Solution to the Puzzle of Human Cardiovascular Disease: Its Primary Cause Is Ascorbate Deficiency Leading to the Deposition of Lipoprotein(A) and Fibrinogen/Fibrin in the Vascular Wall," *Journal of Orthomolecular Medicine* 6 (1991), 125–134.

4. Kåre Berg, "A New Serum Type System in Man—The Lp System," *Acta Pathol.* 59 (1963), 369–382.

5. For further information about these dietary programs, see *The Macrobiotic Way* by Michio Kushi with Stephen Blauer (Wayne, NJ: Avery Publishing Group, 1985); *The New Pritikin Program* by Robert Pritikin (New York: Pocket Books, 1990); *Reversing Heart Disease* by Julian M. Whitaker (New York: Warner Books, 1985); *Reversing Diabetes*, also by Dr. Whitaker (New York: Warner Books, 1987); and *Dr. Dean Ornish's Program for Reversing Heart Disease* by Dean Ornish (New York: Random House, 1990).

6. T.W. Anderson et al., "Ischemic Heart Disease, Water Hardness and Myocardial Magnesium," *Canadian Medical Association Journal* 113 (1975), 199–203.

   M. Speich et al., "Incidences de L'infarctus du Myocarde Sur Les Teneurs en Magnesium Plasmatique Erythrocytaire, Et Cardiaque," *Revue Fr. Endocr. Clin.* 20 (1979), 1550–1562.

7. C.J. Johnson, D.R. Peterson, and E.K. Smith, "Myocardial Tis-

sue Concentration of Magnesium and Potassium in Men Dying Suddenly From Ischemic Heart Disease," *American Journal of Clinical Nutrition* 32 (1979), 967–970.

Barbara Chipperfield and J.R. Chipperfield, "Heart Muscle Magnesium, Potassium and Zinc Concentrations After Sudden Death From Heart Disease," *Lancet* ii (1973), 293–296.

8. G. Behr and P. Burton, "Heart Muscle Magnesium," *Lancet* ii (1973), 450.

   P.C. Elwood, "Magnesium and Calcium in the Myocardium. Cause of Death and Area Differences," *Lancet* ii (1980), 720–722.

   A.S. Abraham, E. Baron, and U. Eylath, "Changes in the Magnesium Content of Tissues Following Myocardial Damage in Rats," *Medical Biology* 59 (1981), 99–103.

9. A.S. Abraham, E. Baron, and U. Eylath, "Changes in the Magnesium Content of Tissues Following Myocardial Damage in Rats."
10. Henrik S. Rasmussen, "Clinical Intervention Studies on Magnesium in Myocardial Infarction," *Magnesium* 8 (1989), 316–325.
11. Y. Rayssiguier, *Hormone Metabol. Res.* 9 (1977), 309–318.
12. M.J. Rowe, J.M.M. Neilson, and M.F. Oliver, "Control of Ventricular Arrhythmias During Myocardial Infarction by Antilypolytic Treatment Using a Nicotinic Acid Analogue," *Lancet* i (1975), 295–300.
13. B.C. Morton et al., "Magnesium Therapy in Acute Myocardial Infarction—A Double-Blind Study," *Magnesium* 3 (1984), 346–352.
14. Henrik S. Rasmussen, "Clinical Intervention Studies on Magnesium in Myocardial Infarction."
15. R.S.A. Tindall, "Cerebrovascular Disease," in Rosenberg, *Neurology* (New York: Grune & Stratton, 1980), 41–77.
16. Bella T. Altura and Burton M. Altura, "Interactions of Magnesium and Potassium on Cerebral Vessels—Aspects in View of Strokes," *Magnesium* 3 (1984), 195–211.
17. Michio Kushi, with Stephen Blauer, *The Macrobiotic Way* (Wayne, NJ: Avery Publishing Group, 1985).

Robert Pritikin, *The New Pritikin Program* (New York: Pocket Books, a division of Simon and Schuster, 1990).

Julian M. Whitaker, *Reversing Diabetes* (New York: Warner Books, 1987).

Julian M. Whitaker, *Reversing Heart Disease* (New York: Warner Books, 1987).

Dean Ornish, *Dr. Dean Ornish's Program for Reversing Heart Disease* (New York: Random House, 1990).

Chapter 6
An Aspirin a Day?

1. H.J. Weiss and L.M. Aledort, "Impaired Platelet/Connective-Tissue Reaction in Man After Aspirin Ingestion," *Lancet* (1967), 495–497.

   J.D. Folts, E.B. Crowell, and G.G. Rowe, "Platelet Aggregation in Partially Obstructed Vessels and Its Elimination With Aspirin," *Circulation* 54 No. 3 (1976), 365–370.
2. C. Patrono et al., "Clinical Pharmacology of Platelet Cyclooxygenase Inhibition," *Circulation* 72 (1985), 1177–1184.

   S. Moncada and J.R. Vane, "Arachadonic Acid Metabolites and the Interactions Between Platelets and Blood Vessel Walls," *The New England Journal of Medicine* 300 (1979), 1142–1147.
3. E.C. Hammond and L. Garfinkel, "Aspirin and Coronary Heart Disease: Findings of a Prospective Study," *British Medical Journal* (1975), 269–271.
4. The Coronary Drug Project Research Group, "Aspirin in Coronary Heart Disease," *Journal of Chronic Diseases* 29 (1976), 625–642.
5. Charles H. Hennekens, Lynne K. Karlson, and Bernard Rosner, "A Case-Control Study of Regular Aspirin Use and Coronary Deaths," *Circulation* 58 (1978), 35–38.

   William T. Friedewald, "Editorial: Aspirin and Coronary Deaths," *Circulation* 58 (1978), 39–40.
6. Aspirin Myocardial Infarction Study Research Group, "A Randomized Trial of Aspirin in Persons Recovered From Myocardial Infarction," *Journal of the American Medical Association* 243 (1980), 661–669.
7. S. Juul-Möller et al., "Double-Blind Trial of Aspirin on Primary

Prevention of Myocardial Infarction in Patients With Stable Angina Pectoris," *Lancet* 340 (1992), 1421–1425.

8. Steering Committee of the Physicians' Health Study Research Group, "Preliminary Report: Findings From the Aspirin Component of the Ongoing Physicians' Health Study," *The New England Journal of Medicine* 318(4) (1988), 262–264.

   Steering Committee of the Physicians' Health Study Research Group, "Final Report on the Aspirin Component of the Ongoing Physicians' Health Study," *The New England Journal of Medicine* 321(3) (1989), 129–135.

9. Steering Committee of the Physicians' Health Study Research Group, "Preliminary Report: Findings From the Aspirin Component of the Ongoing Physicians' Health Study."

10. Steering Committee of the Physicians' Health Study Research Group, "Preliminary Report: Findings From the Aspirin Component of the Ongoing Physicians' Health Study."

    Michael Orme, "Aspirin All Round," *British Medical Journal* 296 (1988), 307–308.

11. Steering Committee of the Physicians' Health Study Research Group, "Preliminary Report: Findings From the Aspirin Component of the Ongoing Physicians' Health Study."

    Steering Committee of the Physicians' Health Study Research Group, "Final Report on the Aspirin Component of the Ongoing Physicians' Health Study."

12. Steering Committee of the Physicians' Health Study Research Group, "Final Report on the Aspirin Component of the Ongoing Physicians' Health Study."

13. Steering Committee of the Physicians' Health Study Research Group, "Preliminary Report: Findings From the Aspirin Component of the Ongoing Physicians' Health Study."

14. Peter Sandercock, "Aspirin for Strokes and Transient Ischemic Attacks," *British Medical Journal* 297 (1988), 995–996.

    M.S. Dennis et al., "Rapid Resolution of Primary Intracerebral Haematoma on Computed Tomograms of the Brain," *British Medical Journal* 295 (1987), 379–381.

15. Steering Committee of the Physicians' Health Study Research Group, "Final Report on the Aspirin Component of the Ongoing Physicians' Health Study."

16. Richard Peto et al., "Randomized Trial of Prophylactic Daily Aspirin in British Male Doctors," *British Journal of Medicine* 296 (1988), 313–316.

Antiplatelet Trialists' Collaboration, "Secondary Prevention of Vascular Disease by Prolonged Antiplatelet Treatment," *British Medical Journal* 296, 320–322.

17. K.A. Fackelman, "Beta-Carotene May Slow Artery Disease," *Science News* 138 (17 November 1990), 308.

18. Daniel Steinberg et al., "Beyond Cholesterol: Modifications of Low-Density Lipoprotein That Increase Its Atherogenicity," *The New England Journal of Medicine* 320 (1989), 915–924.

19. R.A. Riemersma et al., "Risk of Angina Pectoris and Plasma Concentrations of Vitamins A, C and E and Carotene," *Lancet* 337 (1991), 1–5.

20. Herbert Bailey, *Vitamin E, Your Key to a Healthy Heart* (New York: ARC Books, Inc., 1970).

21. Roger J. Williams, *Nutrition Against Disease* (New York: Pitman Publishing Corporation, 1971).

22. Richard Peto et al., "Randomized Trial of Prophylactic Daily Aspirin in British Male Doctors."

23. Steering Committee of the Physicians' Health Study Research Group, "Preliminary Report: Findings From the Aspirin Component of the Ongoing Physicians' Health Study."

Steering Committee of the Physicians' Health Study Research Group, "Final Report on the Aspirin Component of the Ongoing Physicians' Health Study."

24. Aspirin Myocardial Infarction Study Research Group, "A Randomized Trial of Aspirin in Persons Recovered From Myocardial Infarction."

25. K. -H. Grotemeyer, *Thrombosis Research* 63(6) (1991), 587–593.

26. David G. Williams, *Alternatives for the Health Conscious Individual* Vol. 5 No. 5 (November 1993), 37.

K. -H. Grotemeyer, *Fortschr. Neurol. Psychiatr.* 53(9) (1985), 350–353.

27. M.R. England et al., "Magnesium Administration and Dysrhythmias After Cardiac Surgery: A Placebo-Controlled, Dou-

ble-Blind, Randomized Trial," *Journal of the American Medical Association* 268 (1992), 2395–2402.

28. Lloyd T. Iseri, "Magnesium and Cardiac Arrhythmias," *Magnesium* 5 (1986), 111–126.
29. Julian M. Whitaker, *Reversing Heart Disease* (New York: Warner Books, 1985).
30. Richard A. Passwater, *The New Supernutrition* (New York: Pocket Books, a division of Simon and Schuster, 1991).
31. Sheldon Saul Hendler, *The Doctor's Vitamin and Mineral Encyclopedia* (New York: Simon and Schuster, 1990).

Chapter 7
Organizing a Nutritional Therapy Plan and Making It Work

1. Jean A.T. Pennington, *Bowes and Church's Food Values of Portions Commonly Used* (New York: Perennial Library, Harper & Row, 1989).
2. Walter Mertz, "The Essential Trace Elements," *Science* 213 (1981), 1332–1338.
3. Ibid.
4. Herbert Bailey, *Vitamin E, Your Key to a Healthy Heart* (New York: ARC Books, Inc., 1970).
5. Jean A.T. Pennington, *Bowes and Church's Food Values of Portions Commonly Used.*
6. Ibid.
7. Michael P. Ryan, Robert Whang, and W. Yamalis, "Effect of Magnesium Deficiency on Cardiac and Skeletal Muscle Potassium During Dietary Potassium Restriction," *Proceedings of the Society for Experimental Biological Medicine* (1973), 1045–1047.

   Robert Whang and J.K. Aikawa, "Magnesium Deficiency and Refractoriness to Potassium Repletion," *Journal of Chronic Diseases* 30 (1977), 65–68.

   Robert Whang, "Magnesium and Potassium Interrelationships in Cardiac Arrhythmias," *Magnesium* 5 (1986), 127–133.

Glossary

Adenosine triphosphate (ATP). A chemical compound that is essential for producing energy in the body.

Adrenaline. Also known as epinephrine, this is a hormone that stimulates action in the body.

Alpha-tocopherol. The chemical name for vitamin E.

Aluminum. A metallic element implicated in the development of Alzheimer's disease.

Alzheimer's disease. A brain disorder in which the patient becomes progressively less capable of both voluntary and involuntary mental function.

Anaerobic bacteria. Bacteria that live in the absence of oxygen.

Aneurysm. A weak section in a blood vessel, usually the result of a congenital defect, which may rupture and cause a hemorrhage.

Angina pectoris. A condition characterized by pains in the chest, caused by diseases of the heart or blood vessels.

Antioxidant. A chemical agent that inhibits the reaction of a chemical substance with oxygen.

Arrhythmia. *See* Cardiac arrhythmia.

Arteries. Blood vessels that carry blood away from the heart.

Arterioles. Very small arteries that branch off from the arteries and approach the size of capillaries.

Arteriosclerosis. A general term used to describe any condition in which the opening of one or more arteries is physically restricted in some way.

Aspirin. A chemical compound, acetylsalicylic acid, used as a pain-killer.

Atherosclerosis. The gradual blockage of arteries by the formation of fatty deposits on their inner walls.

ATP. Adenosine triphosphate.

Atrioventricular block. A deficiency in the electrical conduction between the heart's atrial and ventricular chambers.

Atrium. Either of the two upper chambers of the heart.

Autotrophic bacteria. Bacteria that thrive in completely inorganic media and need no organic compounds for survival.

Beta-blocker. An agent used to block the chemical sites in the arteries and heart that interact with stimulating hormones, such as those secreted by the adrenal glands.

Beta-carotene. The chemical compound the body uses to make vitamin A.

Blood plasma. The fluid portion of the blood from which the blood cells have been removed.

Blood pressure. The pressure exerted by the blood as it moves through the arteries.

Bufferin. An over-the-counter drug containing aspirin, calcium carbonate, and magnesium oxide.

Calcium. A chemical element found in the body principally in the bones and teeth.

Calcium carbonate. A chemical compound found as the principal component of limestone, marble, and seashells.

Calcium-channel blockers. Chemical species or medications that control the passage of calcium across the cell walls. Examples of these are the drugs Calan, Isoptin, Procardia, Adalat, and Cardizem. Magnesium is a natural calcium-channel blocker.

Capillaries. Tiny blood vessels with walls about one cell thick that allow the exchange of nutrients and wastes between the bloodstream and the body's cells.

Carbon dioxide. A waste product of cell metabolism that is excreted through the lungs.

Cardiologist. A physician who specializes in heart diseases.

Cardiotoxicity. The action or condition of poisoning or causing harm to the heart. The action of the chemical agents that are used to kill cancer cells often cause cardiotoxicity.

Cardizem. A prescription drug used as a calcium-channel blocker.

Catecholamines. Hormones produced by the adrenal glands, such as adrenaline and norepinephrine.

Catheterization. A procedure by which a small tube is inserted into an artery or other natural tubular structure for the purpose of delivering a medication, dye, or tool to a specific site.

Cerebrospinal fluid. The fluid that bathes the spinal column and the brain.

Chelation. A process by which chemical species (molecules or ions) form a molecular coat around a metal ion. This changes the nature of the ion and frequently makes it more soluble.

Chemical free radicals. Atoms, ions, molecules, or fragments of molecules that have at least one electron that is not paired with another electron. In most cases this makes them very unstable and highly reactive.

Chemotherapy. The use of drug therapy. Frequently used to refer to a procedure by which highly toxic drugs are introduced into the bloodstreams of cancer patients for the purpose of killing cancerous cells.

Chlorine. A chemical element used to treat drinking water. Many chemical compounds that contain chlorine are implicated in causing cancer.

Chlorophyll. A large molecule that plants use in photosynthesis and that gives plants their green color. (It is not related to chlorine.)

Cholesterol. A steroid that coexists with fats in the body. It is essential in maintaining the integrity of cell walls.

Cholesteryl esters. Oily substances that are components of lipoprotein molecules. They contain cholesterol chemically combined with fatty acids.

Cold pressor test. A test used to diagnose Raynaud's phenomenon and other disorders that involves immersing a patient's hands in ice water as a means of inducing spasms in the arteries.

Complex carbohydrates. Starches and fibers made up of sugar molecules joined together to form large, complex molecules. Some of them can be broken down to release sugar for the body's use; others cannot. The latter serve as bulk in the intestines.

Congestive heart disease. A condition in which the heart's ability to pump blood is seriously impaired, causing fluid to accumulate in the ankles, lungs, and other tissues.

Coronary angiography. A procedure in which a small catheter, inserted into an artery of the arm or groin, is threaded into the coronary arteries for the purpose of injecting a dye that will reveal arterial blockage when x-ray photographs are taken.

Coronary arteries. The arteries that service the heart.

Cytoplasm. The matter enclosed within a cell's walls.

Deficiency disease. An illness, such as scurvy, caused by the lack of sufficient quantities of an essential nutrient in the body.

Diabetes. A disease in which either the pancreas produces insufficient insulin or the function of the insulin receptors on the cell walls is impaired. It results in reduced metabolism of carbohydrates and enhanced metabolism of fats and proteins.

Diarrhea. A condition in which excessive fluids in the colon cause an abnormally large volume and frequency of excretion.

Diastolic pressure. The minimum pressure the blood flow exerts in the arteries. It is measured between pumping pulses of the left ventricle.

Dietary fiber. The fiber content of food.

Diuretic. Tending to increase the excretion of fluids through the kidneys. Diuretic drugs are routinely prescribed for people with high blood pressure because they decrease the volume of blood and, thus, the pressure it exerts on the blood vessels.

Dolomite. Calcium magnesium carbonate, a form of limestone.

Dyazide. A medication used as a diuretic.

Ectopic heartbeats. Heartbeats that occur as a result of nerve signals that are displaced from their normal positions.

Edema. Accumulation of fluid in the tissues.

EDTA. Ethylenediamine tetraacetic acid, an effective chelating agent.

Electrocardiogram. A recorded trace of the electrical impulses caused by the action of the heart.

Electrolytes. Soluble salts dissolved in the body's fluids. These salts, composed of charged atoms and/or molecules called ions, are the form in which such components as sodium, potassium, calcium, magnesium, zinc, chloride, phosphate, sulfate, and iodide exist in the fluids and cells of the body.

Electron micrographs. Photographs of images viewed in an electron microscope.

Embolus. A loose particle of tissue or a blood clot travelling in the bloodstream. An embolus is capable of blocking the blood's flow at a narrow point in a blood vessel.

Endothelial cells. Flat cells that form the lining of the arteries.

Enzymes. Unique protein molecules that catalyze (bring about) the occurrence of specific biochemical reactions.

Epinephrine. *See* adrenaline.

Esophagus. The tubular passageway from the throat to the stomach.

Extracellular fluids. Fluids that bathe the exterior surfaces of the cells.

Fats. Triglycerides containing mostly saturated fatty acids. Fats are rigid at room temperature, unlike oils, which are also triglycerides, but are liquid at room temperature.

Fatty acids. Organic acids formed when either fat or oil molecules are chemically broken down into their component molecules.

Fibrillation. An arrhythmic condition in which the heartbeats are so rapid that the heart appears to be vibrating rather than beating.

Fibrin. A fibrous protein in blood clots.

Fibrinogen. The protein the body uses to form the clotting material fibrin.

Foam cells. Large cells formed from white blood corpuscles at the site of an injury.

Glucose. The simple sugar the body extracts from food and uses as its main source of energy. It is also known as dextrose.

Hard water. Water with a high mineral content.

HDL. High-density lipoprotein. *See* Lipoproteins.

Heart attack. A condition that occurs when the heart's functions are severely compromised as a result of insufficient blood supply.

Homeostasis. The condition in which the body's processes are operating normally and in balance with one another.

Hypertension. High blood pressure.

Hypertrophy. A condition in which an organ or part of an organ is enlarged (literally, overgrowth).

Infarction. An area of dead tissue that is visible to the naked eye, usually the result of a heart attack or stroke.

Intravenous injection. The procedure in which a liquid is injected directly into a vein.

Invertebrates. Animal species that have no spines.

Ion. An atom or molecule that bears an unbalanced electrical charge.

Ion channels. Microscopic openings in cell walls used for the transport of ions into and out of the cells.

Iron oxide. Iron rust.

Ischemia. The condition of being starved for blood. In the heart or brain, when an ischemic condition becomes severe enough, tissue dies, producing an infarction.

Isomers. Chemical compounds that have the same composition but different arrangements of the component atoms, which gives them different physical and chemical properties.

Kidney stones. Mineral deposits that form in the kidneys and are difficult and extremely painful to pass.

Kilogram. A measure of weight equal to 1,000 grams or approximately 2.2 pounds.

LDL. Low-density lipoprotein. *See* Lipoproteins.

Left atrium. The upper left chamber of the heart. It is the area that receives freshly oxygenated blood from the lungs.

Left ventricle. The lower left chamber of the heart. It propels the blood forcefully into the aorta for distribution throughout the body.

Lesion. Abnormal tissue. Lesions may appear in many organs; in the arteries, they occur in the form of fatty deposits under the arterial lining.

Lipids. Fats and any molecules that dissolve in them. The lipids embrace a group of biochemical substances that are soluble in the same solvents that commonly dissolve animal fats and vegetable oils.

Lipoproteins. Large molecules in which a specific protein is associated with triglycerides, cholesteryl esters, and cholesterol. If the lipoprotein molecule is very large, its density is relatively low, and it is known as a low-density lipoprotein or LDL. When lipoprotein occurs as a relatively small molecule, it is known as high-density lipoprotein or HDL. There are also very-low-density lipoproteins (VLDLs) and intermediate-density lipoproteins (IDLs).

Lumen. The internal dimensions of a tubular structure. In the case of blood vessels, it refers to the free space through which the blood can pass.

Macrobiotic diet. A dietary regimen that is an outgrowth of Asian philosophy and emphasizes balance. Usually it is comprised of 50 to 60 percent whole grains, 20 to 30 percent vegetables, 5 to 10 percent beans and sea vegetables, and 5 to 10 percent soups. Fish is allowed occasionally, but not dairy foods or meat. A variety of different foods provides a balanced diet.

Magnesium. A chemical element essential to life. It controls the activities of other electrolytes and is an essential component of over 300 enzymes.

Magnesium adenosine triphosphate (Mg:ATP). A compound formed by the chemical combination of magnesium with adenosine triphosphate (ATP), this compound is required for the maintenance of normal cellular health.

Magnesium sulfate. A chemical compound, $MgSO_4$, that reacts with water to form Epsom salts.

Methane. The principal component of natural gas.

Mg:ATP. Magnesium adenosine triphosphate.

Microgram. A measure of weight equivalent to one one-millionth (0.000001) of a gram or one one-thousandth (0.001) of a milligram.

Milligram. A measure of weight equivalent to one one-thousandth (0.001) of a gram.

Mineral. An inorganic material that occurs naturally in the earth.

Monocyte. A type of white blood cell.

Multifocal atrial tachycardia. An arrhythmia that occurs when the nerve signals that cause one or both atria to contract are firing off from several locations.

Myocardial infarction. The formation of dead heart tissue that results from a heart attack. It does not occur because the heart is pumping an insufficient amount of blood, but rather because heart tissues are receiving an insufficient amount of blood from the circulatory system.

Myocardium. The muscular tissue of the heart.

Niacin. Known chemically as nicotinic acid; also referred to as vitamin B_3. It is used to lower blood pressure.

Nicotinic acid. *See* niacin.

Nitroglycerin. A medication frequently given to heart patients to dilate blood vessels, increase blood flow, and relieve anginal pain. The proper name is glyceryltrinitrate.

Norepinephrine. A hormone secreted by the adrenal gland. It is related to adrenaline.

Nutrient. An essential food component.

Oils. Triglycerides that are liquid at room temperature. They have unsaturated fatty acids in their chemical structures.

Organic compounds. These include all compounds of carbon (with the exception of carbon oxides, carbonates, cyanides, cyanates, and other compounds of carbon that for convenience are classified as inorganic).

Oxidation. A process by which a substance or material reacts with oxygen, or reacts with another chemical in the same way as it would with oxygen.

Palpitations. A condition in which the heart beats with a noticeable increase in force, whether or not the rhythm is irregular.

Perfusion. The flow of a fluid, such as blood flowing through the vascular system.

Pericardium. The outer covering of the heart.

Phosphorus. An element essential to life. It occurs in phosphate minerals and in organic compounds such as adenosine triphosphate (ATP), which the body needs in order to obtain energy from glucose.

Photosynthesis. The process by which plants convert carbon dioxide into carbohydrates and oxygen.

Placebo. An inert dosage given to the control group in an experiment for the purpose of testing the effectiveness of specific agents.

Plaque. A form of deposit that can accumulate on the interior walls of the arteries and restrict the flow of blood.

Platelets. Components in the blood that are deposited at the site of an injury and assist in the process of clotting.

Potassium. A chemical element that is essential in maintaining the heart's normal rhythm.

Premature beats. Early heartbeats that occur outside the rhythmic pattern of normal beats.

Prinzmetal's angina. Chest pains caused by spasms that occur in the coronary arteries (those that supply the heart's muscles with blood), making them contract in an abnormal way and restricting the flow of blood to the heart. Also referred to as a variant form of angina.

Prostaglandins. Unstable hormonelike substances that are continuously generated. Some prostaglandins inhibit the formation of blood clots in the arteries; others help to form them. Other prostaglandins perform other functions in the body.

Protein. The basic structural material in the muscles, fingernails, and hair.

Raynaud's phenomenon. A circulatory disorder characterized by cold and numbness in the fingers and toes.

RDA. Recommended daily allowance.

Recommended daily allowance (RDA). The amount of a specific nutrient required to avoid dietary deficiency, published by the Food and Nutrition Board, National Research Council of the National Academy of Sciences.

Red blood corpuscles. Blood cells that carry oxygen throughout the body.

Right atrium. The upper right chamber of the heart. It receives spent blood when it returns to the heart from its circuit throughout the body.

Right ventricle. The lower right chamber of the heart. It receives the spent blood from the right atrium and pumps it to the lungs to receive oxygen.

Saturated fats. Fats that contain only saturated fatty acid structures.

Selenium. A trace element that acts as an antioxidant.

Semilunar pockets. Structures inside the veins that help keep the blood flowing toward the heart.

Smooth muscle. In the arteries, muscle that occurs inside the walls and causes the contractions and relaxations we observe in the pulse.

Serum. Blood plasma.

Spasm. An involuntary muscular contraction.

Stenosis. A reduction of the internal dimensions of an artery or other tubular structure.

Stroke. An event that occurs when the blood flow to a portion of the brain becomes so restricted that brain function is severely compromised.

Subendothelial cells. The cells on which the endothelial cells rest and to which they are attached.

Sulfate minerals. Minerals that contain a sulfate ion in their chemical composition. Epsom salts is an example.

Systolic pressure. The maximum blood pressure attained when the left ventricle of the heart contracts and forces blood into the arteries.

Tachycardia. An arrhythmic condition in which the heart beats over 100 beats per minute in its slow mode.

Thrombolytic agent. A chemical agent that dissolves blood clots.

Thrombus. An obstruction in a blood vessel.

Transient ischemic attack. A mild stroke that causes no permanent damage.

Type A personality. A personality that tends to be extroverted and aggressive.

Type B personality. A personality that tends to be introverted and relaxed.

Unsaturated fats. Fats that contain unsaturated fatty acid structures in their chemical composition.

Valence. A chemical term used to express the number of chemical bonds that can be ascribed to a given chemical species.

Veins. The blood vessels that return the blood to the heart.

Ventricle. One of the two lower chambers of the heart.

Venules. Tiny venous blood vessels that connect to, and lead away from, the capillaries.

Vertebrates. Animal species that have spines.

Vitamin. Any one of a number of organic substances that are essential to life. They are present in foods in small amounts.

Vitamin A. A vitamin also known as retinol. It is essential for night vision, among other things.

Vitamin C. A vitamin also known as ascorbic acid or its salts, held by some to be the most powerful antioxidant in the body.

Vitamin E. A vitamin also known as alpha-tocopherol. It is a potent antioxidant in the body.

Warfarin. An anticoagulant used in rat poison and also, in reduced concentrations, for heart patients.

Bibliography

Abraham, A.S., E. Baron, and U. Eylath. "Changes in the Magnesium Content of Tissues Following Myocardial Damage in Rats." *Medical Biology* 59 (1981): 99–103.

Aikawa, J.K. *Magnesium: Its Biological Significance*. Boca Raton, FL: CRC Press, Inc., 1981.

Altura, Burton M. "Ischemic Heart Disease and Magnesium." Magnesium 7 (1988): 57–67.

Altura, Burton M., and Bella T. Altura. "Biochemistry and Pathophysiology of Congestive Heart Failure: Is There a Role for Magnesium?" *Magnesium* 5 (1986): 134–143.

Altura, Bella T., and Burton M. Altura. "Interactions of Magnesium and Potassium on Cerebral Vessels—Aspects in View of Strokes." *Magnesium* 3 (1984): 195–211.

Altura, Burton M., and Bella T. Altura. "New Perspectives on the Role of Magnesium in the Pathophysiology of the Cardiovascular System, II. Experimental Aspects." *Magnesium* 4 (1985): 245–271.

Anast, C.S., and D.W. Gardner. "Magnesium Metabolism." In Coburn Bronner, *Disorders of Mineral Metabolism*. Vol. III. New York: Academic Press, 1981.

Anderson, T.W., L.C. Neri, G. Schreiber, F.D.F. Talbot, and A. Zdrojewski. "Ischemic Heart Disease, Water Hardness and Myocardial Magnesium." *Canadian Medical Association Journal* 113 (1975): 199–203.

Antiplatelet Trialists' Collaboration. "Secondary Prevention of Vascular Disease by Prolonged Antiplatelet Treatment." *British Medical Journal* 296: 320–322.

Aspirin Myocardial Infarction Study Research Group. "A Randomized Trial of Aspirin in Persons Recovered From Myocardial Infarction." *Journal of the American Medical Association* 243 (1980): 661–669.

Bailey, Herbert. *Vitamin E, Your Key to a Healthy Heart*. New York: ARC Books, Inc., 1970.

Behr, G., and P. Burton. "Heart Muscle Magnesium." *Lancet* ii (1973): 450.

Berg, Kåre. "A New Serum Type System in Man—The Lp System." *Acta Pathol.* 59 (1963): 369–382.

Brown, M.J., D.C. Brown, and M.B. Murphy. "Hypokalemia From Beta 2-Receptor Stimulation by Circulating Epinephrine." *The New England Journal of Medicine* 309 (1983): 1414–1419.

Chaudry, I.H., M.G. Clemens, and A.E. Baue. "Alterations in Cell Function With Ischemia and Shock and Their Correction." *Archives of Surgery* 116 (1981): 1309–1317.

Chaudry, I.H., R.N. Stephan, R.E. Dewan, M.G. Clemens, and A.E. Baue. "Use of Magnesium-ATP Following Liver Ischemia." *Magnesium* 7 (1988): 68–77.

Chipperfield, Barbara, and J.R. Chipperfield. "Heart Muscle Magnesium, Potassium and Zinc Concentrations After Sudden Death From Heart Disease." *Lancet* ii (1973): 293–296.

Clark, Luther. "Role of Electrolytes in the Etiology of Alcohol-Induced Hypertension." *Magnesium* 8 (1989): 124–131.

Classen, Hans-Georg. "Magnesium and Potassium Deprivation and Supplementation in Animals and Man: Aspects in View of Intestinal Absorption." *Magnesium* 3 (1984): 257–264.

Cohen, L., and R. Kitzes. "Prompt Termination and/or Prevention of Cold-Pressor-Stimulus-Induced Vasoconstriction of Different Vascular Beds by Magnesium Sulfate in Patients With Prinzmetal's Angina." *Magnesium* 5 (1986): 144–149.

Coronary Drug Project Research Group, The. "Aspirin in Coronary Heart Disease." *Journal of Chronic Diseases* 29 (1976): 625–642.

Dagani, R. "Ion Channels: Discoverers Win Physiology Nobel." *Chemical and Engineering News* 69 No. 41 (14 October 1991): 4, 5.

DeCaestecker, J.S. "The Oesophagus as a Cause of Recurrent Chest Pain: Which Patients Should Be Investigated and Which Tests Should Be Used?" *Lancet* (1985): 1143–1146.

Dennis, M.S., J.M. Bamford, A.J. Molyneux, and C.P. Warlow. "Rapid Resolution of Primary Intracerebral Haematoma on Computed Tomograms of the Brain." *British Medical Journal* 295 (1987): 379–381.

Dhurandhr, R.W., D.L. Watt, M.D. Silver, A.S. Trimble, and A.G. Adelman. "Prinzmetal's Variant Form of Angina With Arteriographic Evidence of Coronary Arterial Spasm." *American Journal of Cardiology* 30 (1972): 9002–9905.

Durlach, J., A. Bara, and A. Guiet-Bara. "Magnesium Levels in Drinking Water and Cardiovascular Risk Factor: A Hypothesis." *Magnesium* 4 (1985): 5–15.

Elwood, P.C. "Magnesium and Calcium in the Myocardium. Cause of Death and Area Differences." *Lancet* ii (1980): 720–722.

England, M.R., George Gordon, Michael Salem, and Bart Chernow. "Magnesium Administration and Dysrhythmias After Cardiac Surgery: A Placebo-Controlled, Double-Blind, Randomized Trial." *Journal of the American Medical Association* 268 (1992): 2395–2402.

Ezzell, C. "Cell Channel Finders Garner Medical Nobel." *Science News* 140 No. 15 (12 October 1991): 231.

Flanagin, A. "JAMA 100 Years Ago: Raynaud's Disease by G.M. Garland, M.D., of Boston." *Journal of the American Medical Association* 262 (1989): 3290.

Flink, E.B. "Magnesium Deficiency—Etiology and Clinical Spectrum." *Acta Med. Scand. Suppl.* 647 (1981): 125–137.

Folts, J.D., E.B. Crowell, and G.G. Rowe. "Platelet Aggregation in Partially Obstructed Vessels and Its Elimination With Aspirin." *Circulation* 54 Number 3 (1976): 365–370.

Franz, K.B. "Magnesium Intake During Pregnancy." *Magnesium* 6 (1987): 18–27.

Friedewald, William T. "Editorial: Aspirin and Coronary Deaths." *Circulation* 58 (1978): 39–40.

Galland, Leo D. "Magnesium in Neuropsychiatric Disorders." *Magnesium* 9 (1990): 324.

Fackelmann, K.A., "Beta-Carotene May Slow Artery Disease." *Science News*, Vol. 138 No. 20 (November 17, 1990).

Gertz, S. David, G. Uretsky, R.S. Wajnberg, N. Navot, and M.S. Gotsman. "Endothelial Cell Damage and Thrombus Formation After Partial Arterial Constriction: Relevance to the Role of Coronary Artery Spasm in the Pathogenesis of Myocardial Infarction." *Circulation* 63 No. 3 (1981): 476–486.

Gertz, S. David, R.S. Wajnberg, A. Kurgan, and G. Uretzky. "Effect of Magnesium Sulfate on Thrombus Formation Following Partial Arterial Constriction: Implications for Coronary Vasospasm." *Magnesium* 6 (1987): 225–235.

Grotemeyer, K. -H. *Thrombosis Research* 63(6) (1991): 587–593.

Gullestad, Lars. "Oral Versus Intravenous Supplementation in Patients With Magnesium Deficiency." *Magnesium* 10 (1991–1992): 11–16.

Hammond, E.C., and L. Garfinkel. "Aspirin and Coronary Heart Disease: Findings of a Prospective Study." *British Medical Journal* (1975): 269–271.

Helfant, R.H. "Hypokalemia and Arrhythmias." *American Journal of Medicine* 80, supplement 4A (1986): 13–22.

Hendler, Sheldon Saul. *The Doctor's Vitamin and Mineral Encyclopedia*. New York: Simon and Schuster, 1990.

Hennekens, Charles H., Lynne K. Karlson, and Bernard Rosner. "A Case-Control Study of Regular Aspirin Use and Coronary Deaths. *Circulation* 58 (1978): 35–38.

Henrotte, J.G. "Type A Behavior and Magnesium Metabolism." *Magnesium* 5 (1986): 201–210.

Higgins, C.B., L. Wexler, J.F. Silverman, and J.S. Schroeder. "Clinical and Arteriographic Features of Prinzmetal's Variant Angina: Documentation of Etiologic Factors." *American Journal of Cardiology* 37 (1976): 831–839.

Horn, Bernard. "Magnesium and the Cardiovascular System." *Magnesium* 6 (1987): 109–111.

Howard, John McLaren, Stephen Davis, and Adrian Hunnisett. "Magnesium and Chronic Fatigue Syndrome." *Lancet* 340 (1992): 426.

Iseri, Lloyd T. "Magnesium and Cardiac Arrhythmias." *Magnesium* 5 (1986): 111–126.

Iseri, Lloyd T., B.J. Allen, and M.A. Brodsky. "Magnesium Therapy of Cardiac Arrhythmias in Critical-Care Medicine." *Magnesium* 8 (1989): 299–306.

Johnson, C.J., D.R. Peterson, and E.K. Smith. "Myocardial Tissue Concentration of Magnesium and Potassium in Men Dying Suddenly From Ischemic Heart Disease." *American Journal of Clinical Nutrition* 32 (1979): 967–970.

Joris, Isabelle, and Guido Majno. "Endothelial Changes Induced by Arterial Spasm." *American Journal of Pathology* 102 (1981): 346–358.

Julius, Stevo, K. Jameson, A. Mejia, L. Krause, N. Schork, and K. Jones. "The Association of Borderline Hypertension With Target Organ Changes and Higher Coronary Risk." *Journal of the American Medical Association* 264(3) (1990): 354–358.

Juul-Möller, S., N. Edvardsson, B. Jahnmatz, A. Rosen, S. Sorensen, and R. Ömblu. "Double-Blind Trial of Aspirin on Primary Prevention of Myocardial Infarction in Patients With Stable Angina Pectoris." *Lancet* 340 (1992): 1421–1425.

Kushi, Michio, with Stephen Blauer. *The Macrobiotic Way*. Wayne, NJ: Avery Publishing Group, 1985.

Landauer, James. "Aspirin and Heart Attacks—A Flawed Study?" *Privileged Information*, 1 June 1988.

Lee, Wai H., and Milton Packer. "Prognostic Importance of Serum Sodium Concentration and Its Modification by Converting-Enzyme Inhibition in Patients With Severe Chronic Heart Failure." *Circulation* 73 (1986): 257–267.

MacLennan, W.J. "Diuretics in the Elderly: How Safe?" *British Medical Journal* 296 (1987): 1551.

Manahan, Stanley E. *Environmental Chemistry*. Boston: Willard Grant Press, 1979.

Marier, J.R. "Magnesium Content of the Food Supply in the Modern-Day World." *Magnesium* 5 (1986): 1–8.

Marier, J.R. "Quantitative Factors Regarding Magnesium Status in the Modern-Day World." *Magnesium* 1 (1982): 3–15.

Marier, J.R. "The Role of Environmental Magnesium in Cardiovascular Diseases." *Magnesium* 1 (1982): 266–276.

Marier, J.R., and L.C. Neri. "Quantifying the Role of Magnesium in the Interrelationship Between Human Mortality/Morbidity and Water Hardness." *Magnesium* 4 (1985): 53–59.

Marier, J.R., L.C. Neri, and T.W. Anderson. "Water Hardness, Human Health and the Importance of Magnesium." National Research Council of Canada Report NRCC—17581 (1979).

Mertz, Walter. "The Essential Trace Elements." *Science* 213 (1981): 1332–1338.

Moncada, S., and J.R. Vane. "Arachadonic Acid Metabolites and the Interactions Between Platelets and Blood Vessel Walls." *The New England Journal of Medicine* 300 (1979): 1142–1147.

Morel, D.W., P.E. DiCorleto, and G.M. Chisolm. "Endothelial and Smooth Muscle Cells Alter Low Density Lipoprotein in Vitro by Free Radical Oxidation." *Arteriosclerosis* 4 (1984): 357–364.

Morton, B.C., F.M. Nair, T.G. Smith, T.G. McKibbon, and W.J. Poznanski. "Magnesium Therapy in Acute Myocardial Infarction—A Double-Blind Study." *Magnesium* 3 (1984): 346–352.

Mountokalakis, Theodore D. "Diuretic-Induced Magnesium Deficiency." *Magnesium* 2 (1983): 57.

Orme, Michael. "Aspirin All Round." *British Medical Journal* 296 (1988): 307–308.

Ornish, Dean. *Dr. Dean Ornish's Program for Reversing Heart Disease.* New York: Random House, 1990.

Packer, Milton, Stephen S. Gottlieb, and Mark A. Blum. "Immediate and Long-Term Pathophysiologic Mechanisms Underlying the Genesis of Sudden Cardiac Death in Patients With Congestive Heart Failure." *American Journal of Medicine* 82, supplement 3A (1967): 4–10.

Packer, Milton, Stephen S. Gottlieb, and Paul D. Kessler. "Hormone-Electrolyte Interactions in the Pathogenesis of Lethal Cardiac Arrhythmias in Patients With Congestive Heart Failure." *American Journal of Medicine* 80, supplement 4a (1986): 23–29.

Parsons, R.S., T.C. Butler, and E.P. Sellars. "Hardness of Local Water Supplies and Mortality From Cardiovascular Disease." *Lancet* iii (1961): 213.

Passwater, Richard A. *The New Supernutrition.* New York: Pocket Books, a division of Simon and Schuster, 1991.

Patrono, C., G. Ciabattoni, and P. Patrignani. "Clinical Pharmacology of Platelet Cyclooxygenase Inhibition." *Circulation* 72 (1985): 1177–1184.

Pennington, Jean A.T. *Bowes and Church's Food Values of Portions Commonly Used.* New York: Perennial Library, Harper & Row, 1989.

Peto, Richard, R. Gray, R. Collins, K. Wheatley, C. Hennekens, K. Jamrozik, C. Warlow, B. Hafner, E. Thompson, S. Norton, J. Gilliland, and Richard Doll. "Randomized Trial of Prophylactic Daily Aspirin in British Male Doctors." *British Journal of Medicine* 296 (1988): 313–316.

Pitkänen, Heljä. "Industrial Possibilities to Interfere With the Salt Problem: Dietary Na/(K+Mg) Ratio." *Magnesium* 1 (1982): 298–303.

Prinzmetal, Myron, R. Kennamer, R. Mewrliss, T. Wada, and N. Bor. "Angina Pectoris I. A Variant Form of Angina Pectoris." *American Journal of Medicine* 27 (1959): 375–388.

Pritikin, Robert. *The New Pritikin Program*. New York: Pocket Books, a division of Simon and Schuster, 1990.

Rasmussen, Henrik S. "Clinical Intervention Studies on Magnesium in Myocardial Infarction." *Magnesium* 8 (1989): 316–325.

Rath, Matthias, and Linus Pauling. "Solution to the Puzzle of Human Cardiovascular Disease: Its Primary Cause Is Ascorbate Deficiency Leading to the Deposition of Lipoprotein(A) and Fibrinogen/Fibrin in the Vascular Wall." *Journal of Orthomolecular Medicine* 6 (1991): 125–134.

Rayssiguier, Y. *Hormone Metabol. Res.* 9 (1977): 309–318.

Regan, T., and P. Ettinger. "Varied Cardiac Abnormalities in Alcoholics." *Alcoholism* 3 (1979): 40–49.

Riemersma, R.A., D.A. Wood, C.C.A. MacIntyre, R.A. Elton, K.F. Gey, and M.F. Oliver. "Risk of Angina Pectoris and Plasma Concentrations of Vitamins A, C and E and Carotene." *Lancet* 337 (1991): 1–5.

Roblin, A. "New Clues to the Power of Magnesium." *Prevention*, April 1989: 33–39.

Rowe, M.J., J.M.M. Neilson, and M.F. Oliver. "Control of Ventricular Arrhythmias During Myocardial Infarction by Antilypolytic Treatment Using a Nicotinic Acid Analogue." *Lancet* i (1975): 295–300.

Ryan, Michael P., Robert Whang, and W. Yamalis. "Effect of Magnesium Deficiency on Cardiac and Skeletal Muscle Potassium During Dietary Potassium Restriction." *Proc. Soc. Exp. Biol. Med.* (1973): 1045–1947.

Sandercock, Peter. "Aspirin for Strokes and Transient Ischemic Attacks." *British Medical Journal* 297 (1988): 995–996.

Schopf, J.W. "The Evolution of the Earliest Cells." *Scientific American* 239 (1978): 110–140.

Schroeder, H.A. "Municipal Drinking Water and Cardiovascular

Death Rates." *Journal of the American Medical Association* 195(2) (1966): 81/125–85/129.

Schroeder, H.A. "Relation Between Mortality From Cardiovascular Disease and Treated Water Supplies." *Journal of the American Medical Association* 172 (1960): 98/1902–104/1908.

Seelig, Mildred S. "Magnesium Requirements in Human Nutrition." *Magnesium Bulletin* 3 (1981): 26–27.

Shils, Maurice E. "Experimental Production of Magnesium Deficiency in Man." *Ann. N.Y. Acad. Sci.* 162 (1969): 847–855.

Siesjö, Bo K. "Calcium and Cell Death." *Magnesium* 8 (1989): 223–237.

Speich, M., B. Bousquet, G. Nicolas, and A.Y. De Lajartre. "Incidences de L'infarctus du Myocarde Sur Les Teneurs en Magnesium Plasmatique Erythrocytaire, et Cardiaque." *Revue Fr. Endocr. Clin.* 20 (1979): 1550–1562.

Steering Committee of the Physicians' Health Study Research Group. "Preliminary Report: Findings From the Aspirin Component of the Ongoing Physicians' Health Study." *The New England Journal of Medicine* 318(4), (1988): 262–264.

Steering Committee of the Physicians' Health Study Research Group. "Final Report on the Aspirin Component of the Ongoing Physicians' Health Study." *The New England Journal of Medicine* 321(3) (1989): 129–135.

Steinberg, Daniel, S. Parthasarathy, T.E. Carew, J.C. Khoo, and J.L. Witzum. "Beyond Cholesterol: Modifications of Low-Density Lipoprotein That Increase Its Atherogenicity." *The New England Journal of Medicine* 320 (1989): 915–924.

Steinbrecher, U.P., S. Parthasarathy, D.S. Leake, J.L. Witztum, and D. Steinberg. "Modification of Low Density Lipoprotein by Endothelial Cells Involves Lipid Peroxidation and Degradation of Low Density Lipoprotein Phospholipids." *Proceedings of the National Academy of Sciences of the United States of America* 83 (1984), 3883–3887.

Stern, Bert, Lawrence D. Shilnick, Gilbert I. Simon, and Harold M. Silverman. *The Pill Book*. 4th ed. New York: Bantam Books, 1990.

Struthers, A.D., R. Whitesmith, and J.L. Reid. "Prior Thiazide Diuretic Treatment Increases Adrenaline-Induced Hypokalemia." *Lancet* Vol. 1 for 1983 (18 June 1983): 1358–1361.

Timmis, Adam. "Modern Treatment of Heart Failure," *British Medical Journal* 297 (1988): 83.

Tindall, R.S.A. "Cerebrovascular Disease." In Rosenberg, *Neurology*. New York: Grune & Stratton, 1980.

Weiss, H.J., and L.M. Aledort. "Impaired Platelet/Connective-Tissue Reaction in Man After Aspirin Ingestion." *Lancet* (1967): 495–497.

Valenzuela, G.J., and L.A. Munson. "Magnesium and Pregnancy." *Magnesium* 6 (1987): 128–135.

Whang, Robert. "Magnesium and Potassium Interrelationships in Cardiac Arrhythmias." *Magnesium* 5 (1986): 127–133.

Whang, Robert, and J.K. Aikawa. "Magnesium Deficiency and Refractoriness to Potassium Repletion." *Journal of Chronic Diseases* 30 (1977): 65–68.

Whitaker, Julian. *Health and Healing*. Vol. 1 No. 2 (September 1991).

Whitaker, Julian M. *Reversing Diabetes*. New York: Warner Books, 1987.

Whitaker, Julian M. *Reversing Heart Disease*. New York: Warner Books, 1985.

Williams, David G. *Alternatives for the Health Conscious Individual* Vol. 5 No. 5 (November 1993).

Williams, Roger J. *Nutrition Against Disease*. New York: Pitman Publishing Corporation, 1971.

Zwada, Edward T., and Nachman Brautbar. "The Possible Role of Magnesium in Hypercalcemic Hypertension." *Magnesium* 3 (1984): 132.

Index